Studies in Systems, Decision and Control

Volume 697

The series "Studies in Systems, Decision and Control" (SSDC) covers both new developments and advances, as well as the state of the art, in the various areas of broadly perceived systems, decision making and control–quickly, up to date and with a high quality. The intent is to cover the theory, applications, and perspectives on the state of the art and future developments relevant to systems, decision making, control, complex processes and related areas, as embedded in the fields of engineering, computer science, physics, economics, social and life sciences, as well as the paradigms and methodologies behind them. The series contains monographs, textbooks, lecture notes and edited volumes in systems, decision making and control spanning the areas of Cyber-Physical Systems, Autonomous Systems, Sensor Networks, Control Systems, Energy Systems, Automotive Systems, Biological Systems, Vehicular Networking and Connected Vehicles, Aerospace Systems, Automation, Manufacturing, Smart Grids, Nonlinear Systems, Power Systems, Robotics, Social Systems, Economic Systems and other. Of particular value to both the contributors and the readership are the short publication timeframe and the worldwide distribution and exposure which enable both a wide and rapid dissemination of research output.

Indexed by SCOPUS, DBLP, WTI Frankfurt eG, zbMATH, SCImago.

All books published in the series are submitted for consideration in Web of Science.

Uriel R. Cukierman

The Art of Teaching, the Magic of Learning: Towards Innovative Higher Education Systems

Springer

Uriel R. Cukierman
UTN-FRBA
Buenos Aires, Argentina

ISSN 2198-4182 ISSN 2198-4190 (electronic)
Studies in Systems, Decision and Control
ISBN 978-3-032-35020-6 ISBN 978-3-032-35021-3 (eBook)
https://doi.org/10.1007/978-3-032-35021-3

Translation from the Spanish language edition: "El arte de enseñar, la magia de aprender: Guía para docentes universitarios innovadores" by Uriel R. Cukierman, © Uriel R. Cukierman 2025. Published by Uriel R. Cukierman. All Rights Reserved.

This Springer imprint is published by the registered company Springer Nature Switzerland AG
The registered company address is: Gewerbestrasse 11, 6330 Cham, Switzerland

To my parents (may they rest in peace), who instilled in me the value of education, and to my children, who are my most precious gift in life.

Foreword

Several years ago, Merieu (2001) pointed out that teaching was a necessary madness. This madness, understood as a transgression of the current canon; as a search for meaning and new horizons in teaching, implied a commitment that was not only didactic, but also ethical and political.

University classrooms (whether virtual or face-to-face) persistently cling to a didactic method characterized by the progressive linear sequence of explanation-application-verification (Litwin, 1997). First explain (theory), then apply through some practice and verify what has been learned with some kind of assessment. Changing it implies, in the first place, feeling a certain pedagogical discomfort: not wanting to teach in the same way we were taught; understanding that the ways of learning have changed; that cultural and digital trends have influenced the ways in which our students (and we ourselves) construct knowledge. These searches often conclude with a type of alteration (for example, the flipped classroom) that allows us to settle into a new pedagogical comfort zone.

This book, however, gives an account of a restless search. Uriel Cukierman invites us to dive into the depths of our teaching practices from his own journey, his history, his trajectory, his questions and his answers. Four decades in which he intertwined teaching, management and research, and gradually generated those horizons of meaning necessary in this era to challenge teaching practices.

When anticipating the reading of this book, trying not to spoil it, I think about how professional careers, in their most visible aspects, but also in their invisible aspects, play unusual tricks on us. As an electronic engineer, Uriel solves problems; he creates, designs, analyses, produces and implements. These skills are also present in his vision of teaching. Knowing the content is not enough; teaching implies a permanent re-design; a creative and artistic invention that is situated and multidimensional (the rupture of pedagogical comfort). The text subtly and charmingly links both professional activities and adds that of research, generously offering glimpses of his years as a researcher and the results of his doctoral thesis. Reflective views, windows that open onto the didactic imagination and theoretical soundness frame these crossroads that the book invites us to travel.

Throughout the book, there are questions about what it means to educate, teach, learn and evaluate in the digital age; it encourages readers to imagine the near future and to delve into the world of artificial intelligence; it offers clues for didactic innovation with technologies (and to take account of the current challenges in university education); it guides and proposes; it invites and points out. Reading is a decision made by the reader. We can read the book sequentially or as a hopscotch game depending on our own interests and concerns. Moreover, the proposal is not only textual. It is nourished by expansions to other non-textual modes that broaden the text and offer the possibility of enriching the exploration. It opens threads that we can pull on in each of the chapters it unfolds.

As a tireless reader, Uriel offers a wide bibliographic framework that can be a starting point for any researcher who wants to delve deeper into these issues. The conceptual contribution of the book is one of its great achievements.

As a restless and generous teacher, he shares concrete challenges that are clues for reviewing and transforming teaching practices at the university.

As a manager for many years, he weaves nuances between macro-, meso- and micropolitics and provides a framework for reviewing institutional decision-making processes at the university.

As a researcher, it invites us not to take anything for granted, to question even the most obvious things and it generates openings towards systematic inquiry, with new questions that deserve attention, consideration and review.

As an engineer, he offers a book that, in its non-linearity, is orderly; it is agile to read, includes very powerful figures and images, and illuminates paths for the understanding of complex issues.

Personally, as the director of his doctoral thesis and having accompanied Uriel on this journey, I see his tenacity as a great quality. It is the tenacity of that "necessary madness"; of that search to change practices that are no longer consistent with the ways in which knowledge is constructed today (Serres, 2013). His pedagogical discomfort with practices that persist even with contemporary cultural and technological trends is a testament to his political and ethical commitment to teaching and research.

The book is an invitation to innovate with meaning; with history; collegially with colleagues. To be part of a collective that is constantly seeking didactic transformation; that dreams of a different and contemporary university; with bridges and interwoven fabrics; with the challenges brought by current cultural and digital trends in complex and uncertain contexts and times.

Are you up for taking up this invitation?

Buenos Aires, Argentina Dr. Carina Lion

References

Litwin, E. (1997). Las configuraciones didácticas. Paidós: Buenos Aires.
Merieu, P. (2001). La opción de educar. Barcelona: Editorial Octaedro.
Serres, M. (2013). Pulgarcita. Buenos Aires: Fondo de Cultura Económica.

Acknowledgments

The chorus of a beautiful song by the Uruguayan singer-songwriter Jorge Drexler describes, in a way that only a good poet can, the essence of life: "Each one gives what he receives/And then receives what he gives/Nothing is simpler/There is no other rule/Nothing is lost/Everything is transformed". The phrase "Nothing is lost, everything is transformed" could mean, for a scientist, the principle of conservation of mass or Lavoisier's law. In my case, it refers to the gratitude I want to express at this moment as I put the finishing touches to this book, because everything I write about here is, in one way or another, what I have received throughout my career as a teacher. Eckhart Tolle, author of *The Power of Now*, says that "it is through gratitude for the present moment that the spiritual dimension of life is opened up".

Reaching this point has not been a matter of chance. My life motto, what I call "the recipe of the three Ps": Passion, Patience and Perseverance, has had a lot to do with it. But I would not have got this far without the support and help of many dear people. I would like to express my gratitude towards them.

First of all, to my parents, who, as I usually say, are still alive in my memory. They instilled the value of education and hard work in me. Then to my family, which we were able to create, with whom I feel "complete", in every sense of the word, and without whom I would not have gotten this far.

I would also like to thank many colleagues from whom I learned through their example. It is difficult to mention some without falling into regrettable forgetfulness, but I will try to do so. I apologize in advance if I fail in the attempt.

To the Virgili family, José María and his son, who encouraged and supported me at the beginning of my teaching career, and while they were still with us.

To my dear friend and teacher Lueny Morell, who transformed my teaching life in a way that I could never have imagined before I met her.

To Guillermo Oliveto, today Dean of the Buenos Aires Regional Faculty of the National Technological University, but, more than anything, a friend from our student activist days at the time of the return of democracy in Argentina. And, on his behalf, to all the colleagues and students of my beloved UTN, my second home for more than 40 years.

To Carina Lion, who very generously directed my doctoral thesis and wrote the loving prologue with which this book begins.

To Telma Souto who, as with my doctoral thesis, dealt with my limited ability to write a text that complies with the rules of good writing and spelling. If there are any errors left in this final text, they are my own and exclusive responsibility.

And finally, to my children, graphic designers, who created the illustrations included in this book and shaped the text you are reading.

I began by quoting a Latin American singer-songwriter and I will end by quoting a female singer-songwriter, the Chilean Violeta Parra, to, in her words, thank life and this teaching profession that has given me laughter and made me cry: "Thanks to life for giving me so much/It has given me laughter and it has given me tears/So I can distinguish happiness from heartbreak/The two materials that form my songs/And your songs which are my own songs/Thanks to life for giving me so much".

Introduction

A vocation is a passion that takes hold of you.
—Santiago Kovadloff

This book is, in a way, the result of a journey that began 43 years ago. My initiation into teaching was completely atypical, one might even say hazardous, although I do not believe much in chance. Borges once said that "what we call chance is our ignorance of the complex machinery of causality" or, as Pasteur commented, "chance favors only the prepared mind" (Rolón, 2023).

From a very young age, I knew I wanted to be an engineer. Therefore, it came as no surprise to anyone in my family when I decided to go to a technical secondary school, where I qualified as an electronics technician. Then, going to the School of Engineering was the logical and expected consequence. What I could not foresee was that, one day, leaving the majestic building of the Engineering School of the UBA[1] (later I went to the UTN,[2] where I graduated as an electronic engineer), I would meet a classmate from high school, who told me that he was working as a teacher in the same middle school where we had met. At that time, I was trying to get some work that would provide me with some money for my personal expenses and, thus, become more independent from my parents. After that casual conversation, I thought that if this former classmate, who had not exactly been one of the best students, was teaching at the middle school, I, who had been a very good student, should be able to do it too. That was how I ended up at my beloved school, then known as E.N.E.T. No. 28 (which we all knew as "el Cuba", because it was located on that street in the Belgrano neighbourhood), and there I met its director, Prof. Nilda Mascali, of whom I had been a student, who offered me, without much ado, the responsibility of the subject Laboratory I. That was how, at the age of just 21 and without any kind of teacher training, I found myself in charge of a course with students who were hardly any older than me. I was far from imagining, at that moment, that education would

[1] Buenos Aires University.

[2] National Technological University.

become my passion and that I would dedicate a large part of my life to this wonderful vocation that is teaching.

Lawyers say that "confession of a part relieves of proof". Indeed, I have just confessed that I was a kind of infiltrator in the teaching profession, and I publicly apologize to all the students who suffered from my lack of training during the first long years of exercising a profession for which I was not prepared. In any case, perhaps it was that "original sin" that later led me to try to train as a teacher: first, learning from the example of my colleagues, who generously supported me in that process; and, later, reading everything I could and attending as many courses and training activities as were available to me. Perhaps due to this personal experience, I have long been promoting the need for all teachers to be trained. In general, in many faculties, teachers are professionals in their disciplines, but they are not education professionals. If, in order to practice any profession, it is necessary to have undergone adequate training, why is it that in many cases, no training is required to practice the teaching profession? It is not a matter of having completed a teaching degree, but rather of at least receiving appropriate training for the task they are going to carry out.

But let's go back to my own journey, which, as I pointed out at the beginning, led me to write this book. Indeed, in this book, you will find much of the experience acquired over more than four decades, the sum of what I have learned from books, courses, colleagues and, more than anything, what my own mistakes have taught me. It was precisely by reading great educators such as John Dewey and Jerome Bruner that I understood the meaning of experiential learning, that is, learning from one's mistakes and building one's own knowledge.

This book contains much of what I have been sharing with colleagues from different parts of Argentina and other countries, and I have tried to write it in the same way that I develop my courses and workshops, with simple and direct language, trying to avoid technicalities and "difficult" words. This book aims to help those who, like me many years ago, want to teach but haven't had the opportunity or time to participate in specific training programs. I have written it from the first-person singular and towards the second person singular or plural, from me to you or to all of you. It is a text that attempts to answer many of the questions that teachers may have regarding the exercise of the profession; therefore, you will find a series of questions that are most likely the same ones that you have asked yourself at some time. In an attempt to make this book a living resource, you are invited to send me any other questions that are not covered here. I promise to do my best, using my knowledge and the resources available to me, to try to answer them and thus continue to improve and expand this text. And, perhaps, after some time, to produce a second edition, expanded and augmented with contributions from all of you, dear colleagues.

Chapter 1 includes a few basic questions and answers about education and, in a way, it presents my own approach to the subject, which does not pretend to be original or new, but, as I said before, is the result of my own experience.

In Chap. 2, I focus on teaching and learning, two issues that are closely related, but that do not always depend on each other; sometimes, there is teaching, but there is no learning; on other occasions, there is learning even though there is no teaching; and, sometimes, there is neither, even though there are teachers and students. Surely, those who have been in the education sector for many years can imagine or remember situations that reflect any of these scenarios. In addition, I describe different types of learning or ways of learning.

Chapter 3 deals with assessment, the big issue that concerns us all (students, teachers, institutions and the State) and that generates so much controversy. In general, the issue is often underestimated when it is equated with exams and their grades, but assessment is much more than that. Assessment allows each of the actors involved in education to know to what extent things are being done well or what needs to be improved; and, in this approach, my focus is already taking shape. Therefore, I analyse the types of assessments and try to answer what, who and how to assess.

To achieve anything, we first have to think about how we are going to do it. From something as simple as traveling to a place to deciding what we want to be as people, everything requires a minimum of prior planning. In education, this planning is carried out when designing a curriculum, when planning a course or when preparing a class. Chapter 4 deals with this issue, which is often resolved in a merely formal way to comply with administrative requirements, but which, in reality, is the basis of effective academic organization. If this process is carried out lightly, the results obtained are likely to be poor or unexpected.

In Chap. 5, I tackle an issue that, in my opinion, is of transcendental importance; even more so, in today's world crossed by social networks, robots, artificial intelligence and other developments typical of modernity. It is about thought and, in particular, critical thinking. In this era, in which we live subjected to the effect of algorithms that analyse our behaviour to profile the information we access, it is extremely important to learn to analyse and evaluate biases and intentionality in the information that comes to us on a continuous basis. In this chapter, I analyse some issues that, in my opinion, should be part of all curricula.

Nowadays, the issue of technology in education is unavoidable. In reality, it always was, as books, chalk, blackboards and other analogue tools are technologies that have been used since earliest times. But there is no doubt that digitalization has changed the role of technology in education. Ever since the appearance of digital calculators, but even more so with computing and the Internet, technology has become a central factor in the design and implementation of educational proposals. Although the subject is too broad to be analysed in its entirety, in Chap. 6 I focus on some issues that I consider to be more significant today.

Although it is a technology and, therefore, should perhaps have been part of the previous chapter, given its current and future importance, I decided to dedicate an entire chapter to the subject of artificial intelligence. Today, we are overwhelmed by hundreds or thousands of articles on the subject from the most diverse origins and approaches. Chapter 7 offers an introduction for those who want to have a basic understanding of the subject, as well as some considerations regarding issues that not only teachers should take into account, but that it would also be advisable for our students to consider when using this tool.

Chapter 8 is about change because I understand that there is no learning without change, but change is not always easy or automatic. I analyse change at the institutional, teaching and student levels and also the what, why and for what purpose of change. There are methods to promote and facilitate change, and it is convenient to use them to achieve the expected or desired results.

Finally, Chap. 9 presents a series of challenges that I believe education faces today (with a particular focus on higher education) and which, in my opinion, should be analysed and considered at all levels of management to keep educational institutions in line with the demands of contemporary society. Educational institutions fulfil a fundamental social role and should be at the service of the society that sustains them, updating their practices and purposes, since, if they do not do so, they will cease to be relevant in the face of offers that only pursue profit or particular interests.

I would like to make an important terminological clarification here. Throughout the text, I will use the words pupil and student interchangeably, since, according to the Dictionary of the Spanish Language, pupil means 'disciple, in relation to a teacher, of the subject being learned or of the school, college or university where one studies' and student means 'a person who studies in an educational establishment'. (Real Academia Española)[3]; that is, they are equivalent terms or directly synonymous. And, since we are dealing with semantic issues, it is worth clarifying that, contrary to a fairly widespread assumption, the word pupil ("alumno" in Spanish) does not mean 'without light' (a = 'without' and *lumnus* = 'light'), since *lumnus* does not exist in that form in Latin. Instead, this word comes from the Latin *alumnus* which, in turn, derives from the term alĕre which means 'to feed'.

It is worth mentioning, to conclude this introduction, that although the chapters are in sequential order, it is not mandatory to read them all or in that order. I sincerely hope that my teaching colleagues will find some useful ideas or resources in this book; and if that happens, I will be satisfied. In any case, I will appreciate all the opinions, criticisms, suggestions or recommendations that you send me. After more than 40 years as a teacher, I am still learning and, hopefully, I can continue to do so for as long as my reasoning allows me to. Thank you for your interest in my work.

A Note on Translation

As you read, please keep in mind that this work was developed in Argentina and originally written in Spanish. Consequently, you will encounter references to Argentine

[3] TN: This book is an English translation of the original Spanish version.

sources and local contexts that have been kept intact to honour the spirit of the original text. For the supplementary YouTube videos included, don't worry if you don't speak Spanish—you can simply activate the automatic translation feature within the video player's closed caption settings to access English subtitles.

Contents

Chapter 1
Questions and Answers about Education

Education, in its broadest sense, is the means of this social continuity of life.
John Dewey

We all talk about education, but we do not always have a clear idea of what we mean by that word. Politicians, leaders, union representatives, parents, and teachers all speak about education, but for each of them, education may mean different things. A person is said to be "educated" when they behave according to certain general norms of civility and courtesy. How many times have we heard someone being called "uneducated"[1] when they act in an antisocial or inconsiderate way? When someone has a special inclination or liking for art, we say they have an artistic education. Moreover, there is formal and informal education, public and private education, paid and free education, etc. For this reason, before moving forward with other topics, I find it necessary to clarify what I mean by education in this book. I will refer specifically to formal education, regardless of its management system or funding source, with a particular emphasis on higher education institutions. However, much of what I describe here is also perfectly applicable to other levels, such as primary and secondary education.

With these clarifications in place, I will attempt to answer some questions about education (see Fig. 1.1).

In all these questions, I have included the word change because I understand that education implies changes in the person being educated, and because education requires changes in institutions and their stakeholders, which change so that education improves (I will address the topic of change management in a later chapter). Unfortunately, educational environments tend to be very conservative.

[1] TN: The original term in Spanish is "maleducado" which means ill-mannered.

TN: Translation of the book "The Art of Teaching, the Magic of Learning: Guide for innovative university professors" authored by Uriel Cukierman and recently published in Spanish in Amazon https://www.amazon.com/dp/B0DTQ5RC2B.

U. R. Cukierman, *The Art of Teaching, the Magic of Learning: Towards Innovative Higher Education Systems*, Studies in Systems, Decision and Control 697,
https://doi.org/10.1007/978-3-032-35021-3_1

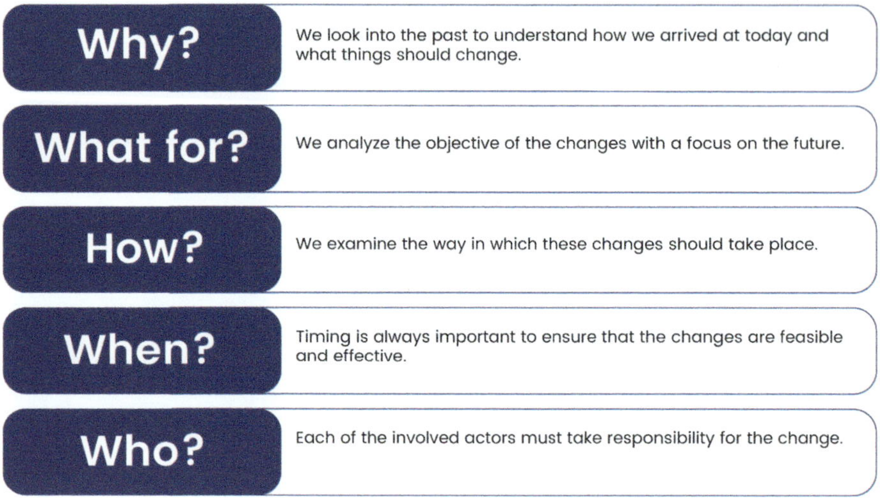

Fig. 1.1 Questions about education

In a talk I gave at the Argentine Center of Engineers back in 2015, I described a person who had fallen asleep in 1915 and, upon waking up a hundred years later, felt utterly lost because everything was different: buildings, cars, communication media, and so on. Everything had changed so much that they could not recognize anything around them—until they arrived at university. There, they felt reassured because everything was just as it had been back then.

On YouTube, there is a video that describes this situation very eloquently. I recommend watching it:

https://bit.ly/juicio_educacion

"The university classroom [...] is facing, for the first time in history, conditions that are stripping it of the meaning it once had. The traditional class as we used to conduct no longer makes sense." (Maggio 2018, p. 18). Can we then continue pretending that nothing has changed? In other words, can we keep doing the same things we have always done and believe that this benefits our students? My answer is that we can no longer afford to bury our heads in the sand and ignore what is happening. Let us begin by analysing in detail the questions I previously posed as a first step toward understanding what is happening and how we can change it.

1.1 Why?

According to a well-known historian and philosopher, "history is not just the study of the past; it is the study of change." (Harari 2023a). Furthermore, in his book 21 Lessons for the twenty-first Century, the chapter on education is titled "Change is the Only Constant" (Harari 2018, p. 285). Thus, it is relevant to briefly examine the history of education. In a broad sense, education is intrinsic to living beings. However, it was not until the Industrial Revolution in the mid-eighteenth century that education came to be understood as a production chain. The school, as we know it today, is a result of that logic.

> In the middle of town there is a large concrete building divided into many identical rooms, each room equipped with rows of desks and chairs. At the sound of a bell, you go to one of these rooms together with 30 other kids who were all born the same year as you. Every hour some grown-up walks in and starts talking. They are all paid to do so by the government. One of them tells you about the shape of the Earth, another tells you about the human past, and a third tells you about the human body. (Harari 2018, p. 292)

Any resemblance to reality is not purely coincidental.

I also recommend watching a YouTube video based on a talk by Ken Robinson:

https://bit.ly/video_KR

After this brief historical review, the need for change should be clear. Several industrial revolutions have occurred since the current educational system was created; the world has changed and continues to change at a dizzying pace. Today, access to information is virtually unrestricted and unlimited (at least in free and democratic societies). Technology is ubiquitous and omnipresent, and its most recent expression is generative artificial intelligence, whose public emergence just over a year ago[2] has sparked such a revolution that some wonder whether we are witnessing the dawn of a new type of social organization where machines control our daily lives. In my humble opinion, to prevent this "fifth horseman of the apocalypse" from taking over humanity, a series of radical changes is essential: one of them (and perhaps the most important) being education. We must stop planning the future by looking in the rearview mirror and start making the changes that future generations demand.

---

[2] Regarding the date on which this book is being written.

1.2 What for?

The next question I pose is: what should we change for?? What I stated in the previous section is certainly a strong indication. However, it is not enough to say that we need to change simply because what we were doing before no longer works. We must conceive and implement a new education system that serves to shape the society of the future. And when I say "future," I am not referring to the year 2100, I mean tomorrow, next year. It is well known that, regardless of how much goodwill is put into the process, changes in education do not happen instantly. However, we also cannot afford to wait for an entire generation to see the results we seek. Changes must occur simultaneously at all educational levels. As I have explained earlier, this book focuses on higher education. Therefore, I will leave matters related to prior educational levels to experts in those fields and analyse what the objectives of change in higher education should be.

I could conduct a discipline-specific analysis, but some issues are entirely transversal. One of them is the need to foster the development of critical thinking in young people (and in adults too, but I will leave that discussion for another time). Regardless of the profession our students aspire to, being able to think critically is now essential for life. Unfortunately, this competency is not included in nearly any current curriculum (and I say "nearly" because I hesitate to claim "none" outright). For this reason, I will dedicate an entire chapter to this topic later in this book.

Another equally generic and discipline-independent issue is the need to create flexible learning paths, allowing students to choose what they want to learn. It is true that in professional degree programs such as Engineering, Law, and Medicine, there are certain unavoidable requirements to certify graduates' professional competencies. However, today's curricula are overloaded with content that, if not already obsolete, will be in the near future. We must focus curricula on fundamental concepts that will remain relevant over time and on the development of transversal or generic competencies, which are essential not only for professional performance but also for life in modern society, such as creativity, problem-solving, collaboration, and the ability to learn how to learn (Furman 2021, p. 19).

1.3 How?

I have already touched upon some aspects of the how in previous paragraphs, but at least two key issues are essential. One must come "from the top down", meaning that educational institutions' leadership must promote a culture of innovation and change. The other must come "from the bottom up", meaning that change depends on each of us, including students and faculty, who must start implementing changes in classrooms, even if, at the institutional level, no significant changes have yet taken place. I mention students because, in a "student-centered learning" approach (Cukierman 2018), a topic I will revisit throughout this book, they must take the lead, not only in the classroom but also in university governance, demanding and, at the same time, contributing to the necessary changes. After all, they will be the first to benefit from receiving an education suited to contemporary realities. As educators, we must recognize that our role is not to repeat what has been done for generations; we must dare to innovate in the classroom. Doing so may lead to mistakes, and we must not fear that possibility. After all, we are not heart surgeons operating on live patients, not because our work is any less important, but because, without a doubt, our mistakes are easier to rectify. Innovation inevitably involves the risk of error. Steve Jobs, one of the greatest innovators of recent decades, used to say: "Sometimes when you innovate, you make mistakes. It is best to admit them quickly and get on with improving your other innovations." (Elevate Society). Throughout this book, I will share ideas and concrete proposals to help initiate this path of innovation.

Another essential element in determining how to enact change is faculty training. This training should be both provided and required by institutions, as well as sought out and accepted by educators themselves. In any profession, continuous professional development is essential to remain aligned with job market demands. Teaching is no different. Although I will discuss this topic further later, it is important to note that teaching is a profession and, as such, requires proper training. Good intentions are necessary, but they are not sufficient. In a 2023 survey of over 1000 higher education faculty members from different institutions in 46 countries, 96% responded that pedagogical training should be offered and/or required (see Fig. 1.2).

Another essential factor that, though obvious, must still be mentioned is recognizing those who carry out their work with dedication, commitment, and effectiveness. A professor who receives poor evaluations in surveys or whose courses are not passed by the majority of students should not be considered in the same way as another who obtains good results in both indicators. Such recognition should be explicit and

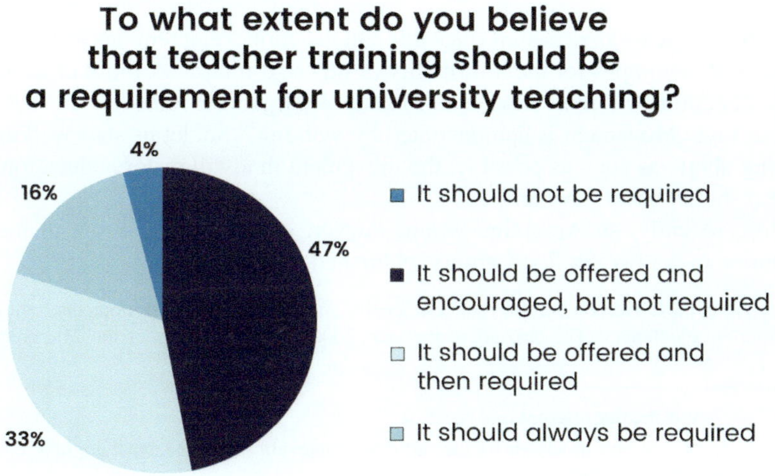

Fig. 1.2 About faculty training in university

should impact faculty career progression as well as salary. I know this is a controversial proposal, but that is how the real world works. In professional life, people advance in rank and salary based on merit, not just on the mere passage of time (i.e., "seniority"). If we are to implement the changes that education urgently needs, these issues must at least be discussed and seriously evaluated by the entire academic community. The periodic evaluation of faculty positions and competitive hiring processes were already established in the 1918 University Reform (over 100 years ago!).[3] However, today we see how some faculty positions have become lifetime appointments, and competitive selection processes have, in many cases, turned into bureaucratic formalities, largely due to a misguided approach to unionism. I firmly believe that devaluing the teaching profession does not protect it. While demanding better salaries, we must also promote a profession of excellence that rises to the challenges of our time.[4]

1.4 When?

Is there any answer to this question other than NOW? Just a few decades ago, Argentina was a model country in terms of its educational system. Today, unfortunately, we have dramatically regressed. All kinds of indicators confirm this decline. If we do not change the present, we are mortgaging the future.[5] And here the memory arises of the many times repeated, but rarely accepted words of the renowned Spanish philosopher, who already in 1939 said:

> Argentinians, focus on what truly matters! Stop with personal squabbles, suspicions, and narcissism. You do not realize the tremendous leap forward this country will take the day its people bravely decide to engage directly with reality, to deal with and concern themselves with actual issues, instead of constantly being on the defensive, having their brilliant intellectual potential stifled by personal disputes. (Ortega y Gasset 1995, p. 99)

More than 80 years have passed since this call to action, yet it remains just as relevant today. With apologies for the informality of this next reference, those of us who are now older adults will remember an unforgettable Argentine comedian and actor who used to say: "Movement is demonstrated by walking." So, let us start walking now to bring about, as soon as possible, the movement that will restore education to the place it should never have left.

More recently, an Argentine writer, lawyer, and politician wrote in his book Argentina in the twenty-first Century, published in 1985:

> Argentina must overcome its current deficiencies and prepare the generations that will face imminent challenges. The citizens of the year 2000, the first leaders of the 21st century,

---

[3] TN: This is specifically referred to Argentina.

[4] TN: All this paragraph was written for the particular context of Argentina, but it might be applicable in a lot of different countries.

[5] TN: Even though this comment is specifically referred to Argentina, it might also be applicable in several countries.

> are, let us remind ourselves of this, the same ones who are now learning the alphabet. For Argentina to move forward in this emerging world, the time to start is now. (Terragno 1985)

Almost 40 years have passed since that "now." We cannot delay any longer.

1.5 Who?

Finally, we arrive at the last but no less important question. And once again, is there any answer other than EVERYONE? Although I have previously mentioned some key actors within academia, this issue concerns everyone, from the highest national authorities to ordinary citizens, each with their own level of responsibility. No one can look the other way, because in this mission, our future as a society is at stake.

Chapter 2
About Teaching and Learning

Teaching is the art of presenting knowledge in a way that is accessible and meaningful to students.
Jerome Bruner

What does it mean to teach and to learn? Can we say that a teacher is really teaching when they just talk for hours in front of a classroom? Can we say that a student has learned when they repeat in an exam what the teacher has said? We probably all have some way of defining these words, but in this chapter, I intend to analyse the usual definitions in greater depth and contextualize them in terms of the title I proposed for this book: "The art of teaching, the magic of learning".

The word art,[1] in general, is associated with the artistic. In fact, in the Dictionary of the Spanish Language, the second definition of this word is "manifestation of human activity through which reality is interpreted or the imagined is shaped with plastic, linguistic or sound resources" and the first definition is "capacity, ability to do something" (Real Academia Española). In addition, the following synonyms, among others, are mentioned: talent, genius, faculty, aptitude, condition, disposition, inspiration, dexterity, skill, practice, experience, mastery. Taking these meanings into account, who could say that teaching is not an art? According to the aforementioned definition, in order to develop this art, it is necessary to have the capacity or the ability to do so, which basically implies having three elements (see Fig. 2.1):

Vocation: We could say that it is about "wanting to do it".
Dedication: It could be assimilated to the commitment required to develop it in an appropriate way.
Effort: Because it is difficult to obtain something valuable in life without making an effort to achieve it.

Let's now turn to the word teach. What does it mean to teach? Some might say it means to transmit or transfer knowledge, but this idea is not, in my opinion, at all appropriate.

[1] TN: "Arte" in the original text in Spanish.

U. R. Cukierman, *The Art of Teaching, the Magic of Learning: Towards Innovative Higher Education Systems*, Studies in Systems, Decision and Control 697,
https://doi.org/10.1007/978-3-032-35021-3_2

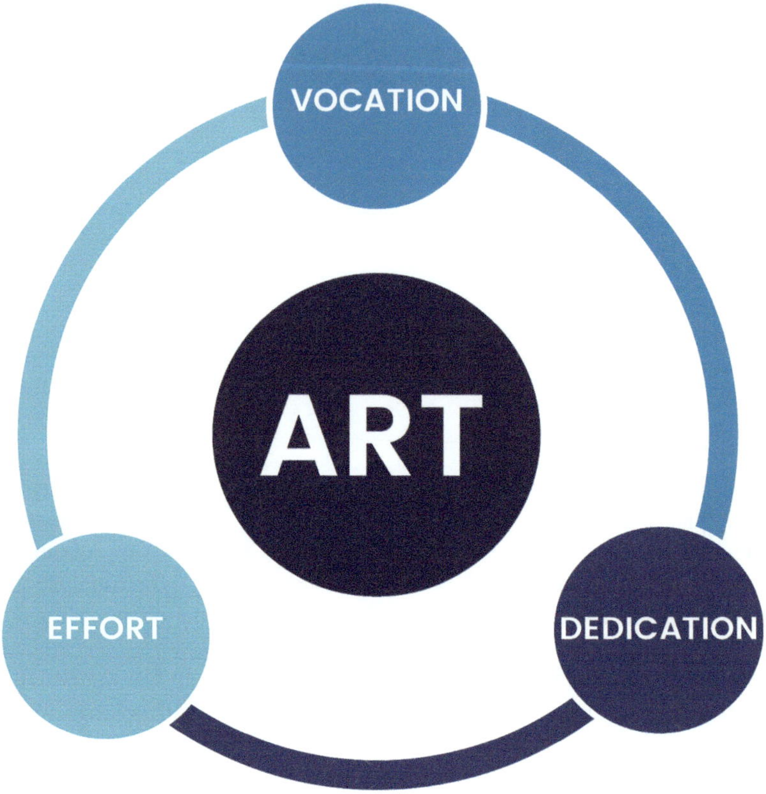

Fig. 2.1 Elements involved in art

> University education is generally perceived as the transmission of information, and teaching and assessment systems around the world are based on this assumption. Classrooms and teaching materials are specifically designed for one-way. (Biggs 2005, p. 41)

What Biggs proposes is evident in the physical layout of classrooms, with the teacher at the front and the students sitting in the same way as in a cinema or theatre, as mere spectators. Paulo Freire expresses himself in the same sense when he says: "Teaching is not transferring knowledge but creating the possibilities for its own production or construction" (Freire 1997, p. 47). On the other hand, it is interesting to know that, in Greek, to teach is *didásko* (διδάσκω), also related to the term *didactics*. There is consensus that Comenius was the father of didactics. It is precisely in his *Didactica Magna* that he talks about the "art of teaching" (Comenio 1998, p. 36). But, to effectively develop this art of teaching, it is not enough to know the subject: "… to teach well we need to know a lot about what we teach (the content), and also a lot about how to teach that particular content to our group of students (the specific didactics)" (Furman 2021, p. 94).

However, especially in higher education, to be a teacher, knowing what we teach is enough. In fact, Law No. 24521 on Higher Education in force in Argentina (InfoLEG 1995) states in article 36: "Teachers of all categories must hold a university degree of the same or a higher level than that in which they teach"; that is to say, in order to teach, no teacher training is required. This is, in my opinion, at least, striking. To be a lawyer, accountant, engineer, etc., you need to have the degree corresponding to that profession, but it seems that teaching is not a profession, since no specific training is required to practice it. In fact, as I mentioned in the introduction, I myself worked for many years as a teacher at secondary and university level without having any training to do so; I did it, as many do, emulating those who were my exemplary teachers. That was the case until I realized that, with that, it was not enough, that I needed to learn to teach or, as I like to say, learn the best way to make my students learn. I know that this may make many of my university teaching colleagues uncomfortable, but I believe that this should change and that, in order to be a teacher at any level, specific training should be required (and offered). It is not a question of us having to have a doctorate in education or of us having to be pedagogues, but we do need to have pedagogical training in line with the tasks we are going to perform; and that training, as in any other discipline, should be continuous or, at least, periodically updated to adapt to the changes that occur in society and in our students.

It is interesting at this point to bring up the approach proposed by Alliaud, who refers to teachers as "artisans of teaching". The author characterizes the act of teaching as "a kind of particular vibration carried by teachers". And she describes what she calls "trade secrets", which are those that "take practical form as they are taught, and seem to be the product of a rare formula of methods, techniques, ways of acting that in theory allow us to distinguish what works well from what does not". (Alliaud 2017, p. 41).

Returning to the art of teaching, I would like to quote Einstein here (I will do so repeatedly throughout this book, as I am passionate about his life and work), when he says: "It is the supreme art of the teacher to awaken joy in creative expression and knowledge". (Calaprice, The ultimate quotable Einstein 2011).

Now we will analyse the word *learn*. Many times, we have heard or read texts in which the words *teach* and *learn* are linked, as if one could not exist without the other. Therefore, I think it is necessary to point out that this cause-effect relationship is neither linear nor immediate; that is to say, it is possible for someone to teach but not get their students to learn (in which case, one would have to doubt whether that person is really teaching); or, on the contrary, for a person to learn without needing to be taught, the so-called "autodidacts". In my courses for teachers, I often use a comic strip by the American illustrator Bud Blake in which a boy tells a friend that he has taught his dog to whistle. Surprised, the friend says he can't hear it whistling, to which the boy replies that he said he taught it, not that the dog learned to do it itself.

Going back to the academic environment, how many times have we heard colleagues say, in a very casual manner, that they have given a lesson and that, if the students did not learn, it's their problem. There are also courses in which the majority of students systematically fail, and the teacher does not take responsibility for the results, as if those students had conspired not to learn. For this reason, the second part of the title of this book refers to the magic of learning. I will expand on this concept a little.

The word *magic* has many meanings depending on the context in which it is used, but, in general, it is associated with something that is wonderful or surprising. In that sense, I wonder if there is anything more wonderful than learning. Isn't it "magical" to see how babies learn to eat, walk and talk, and all the skills they acquire as they grow up? Think of the joy we probably felt as children or teenagers learning to ride a bike, swim or play a musical instrument. There will be those who do not have good memories of learning these skills, but, surely, in those cases, it was because it was made compulsory or forced in some way. On the other hand, doing an etymological analysis of the word *magic*, we also find references that give the word the meaning of 'being capable' or 'having the capacity' (Harper).

I wonder, then, what learning is. Again, turning to the Dictionary of the Spanish Language, the first definition is 'to acquire knowledge of something through study or experience'. (Real Academia Española). However, this definition talks about "acquiring knowledge", which does not sit very well with the constructivist theories on which we currently rely, which propose that learning occurs when the individual "constructs" knowledge as a result of their experience in the world; that is, learning is, essentially, active. As I mentioned in the previous paragraph, the most basic and important learning in life takes place in an active, experiential way; and it is incorporated so well that we do not lose it for the rest of our lives. One can stop cycling or swimming for years, but when one tries again, one does so without major problems. This is the "magic" of true learning; and for this learning to occur, there must be teachers who have mastered the "art" of teaching.

At this point I would like to suggest that you (yes, you, the person reading this book) carry out an activity that will only take a minute.

Activity 2.1: What is the first response that comes to mind when you read the following question?

What is more important? What we teach or what our students learn?

What did you answer? If you answered that the most important thing is what we teach, I'll ask the question in a different way:

If our students don't learn, what's the point of what we teach them?

This raises several additional questions:

- How do we know if our students have learned?
- Do we know for sure what we want them to learn?

- Is what we want them to learn really valuable or useful for their life or their profession?

Answering these questions will take us into the field of assessment, a subject I will develop in the next chapter. Let's go back to the subject we were dealing with previously. I propose another short activity.

➡ Activity 2.2: Write down what learning is or what learning represents for you.

Have you done it yet? Go on, no cheating ☺. Close the book for a moment and complete the activity before you read on.

Now let's describe some of the characteristics of learning. See if anything you wrote matches or resembles this:

1. Learning is a **process**, not a product.
2. Learning involves **changes** in knowledge, beliefs, behaviour or attitudes
3. Learning is not something that is done **to** students, but something that **students do for themselves.**

The **process** I refer to in the first point is something that happens in the minds of our students, but we can infer whether it has occurred through their products or performances; that is, from what they do with what they have learned, how they apply it in the resolution of a particular situation or problem. This is where the importance of the **problem or project based learning** method, which I will refer to later in this chapter, comes in.

The **changes** mentioned in the second point do not happen instantly, they take time; and that time is not the same for all students nor does it happen in the same way. Therefore, it is very important not to follow the pace of those who are able to make those changes more quickly. A good strategy for dealing with these differences is **peer learning**; that is, those who learn faster help those who need more time. In this way, the former reinforce their learning and the others feel more supported. Let's banish the idea that the fastest learner is the best student or, even worse, the smartest. The important thing is not the speed, but the quality of learning.

Last but not least, what is said in the third point confirms the importance of **student-centered learning**, a concept to which I will also return in detail later. To such an extent that it cannot be said that one person "teaches another". Learning is something that each individual does for themselves and is the direct result of how that individual responds to their own experiences, whether conscious or unconscious, past or present (Ambrose et al. 2010, p. 3). As the renowned French pedagogue says: "Pedagogy must defend that every being is capable of being educated, can learn and grow, but it is the being itself that learns and grows, and I cannot learn and grow in its place" (Meirieu 2022).

There are many definitions of learning, but one that I find particularly appropriate, in line with what I have been saying, is the one expressed by Herbert A. Simon, one of the founders of the field of cognitive science and a Nobel Prize winner, who said:

"Learning is the result of what the student does and reflects on and only of what they themselves do and think. The teacher can improve learning only by influencing what the student does to learn" quoted in (Ambrose et al. 2010, p. 1).

Beyond the definition of learning that we adopt, there are infinite ways of learning. One could say that there are as many as there are students or, perhaps, as many as there are people, because it is not only the student who learns. In fact, we learn all our lives, from the moment we are born; and only for part of that life do we do so in a school. Various theories have been developed about "learning styles" (Felder and Silverman 1988; Kolb 1976; Fleming and Mills 1992) and, as theories, they have their advocates and their detractors. It is not my intention to discuss them in this text, but it cannot be denied that each person learns in different ways and at different times. Therefore, it is our responsibility as teachers to propose different learning strategies to allow for true inclusion. Once, many years ago, I heard a colleague, Luis Osin,[2] say at a conference that **true democracy in education is not about giving everyone the same thing, but about giving each person what they need**.

How many times have we heard colleagues in the staff room complain about students who, despite having repeatedly explained the subject, still do not understand? The problem is not solved by explaining the same thing over and over again, but by proposing alternative ways of learning; for example, by proposing an approach from a different perspective or having it explained by a classmate who did understand it.

What follows is a proposal for different learning strategies that can be used according to the different scenarios, circumstances and contexts in which we find ourselves teaching. I have already mentioned some of them in this text.

2.1 Active Learning

Actually, this term encompasses all the methods listed below. In short, learning can be considered active when students carry out any activity related to the course that is not simply listening to the class and taking notes (Felder and Brent 2016, p. 113). I will now detail some strategies that are as simple as they are effective.

Think-Pair-Share (TPS): is a very useful strategy to use when we have to explain a topic for a long time, which can lead students to lose attention and become distracted. It involves asking a question about something that has not yet been explained and asking them to think (silently and for one minute) what their answer to the question would be. It is important to let them know that there are no wrong answers at this stage. In the second minute, they will be asked to share and discuss the answer they thought of with the person next to them (in pairs). Finally, some of the pairs will be

[2] https://math.escaque.org/bio/Luis_Osin/Luis_Osin_sp.php.

asked to share their answers with everyone, thus generating a group discussion about which answer might be the most correct; the class will then resume. This activity can be repeated as many times as necessary throughout a class.

One-minute report: this activity is appropriate for the end of the class, as it makes students mentally recap what they have seen, trying to focus on what they thought was most important or worth remembering. To do this, they are asked to write down the three issues they found most significant in the class on a piece of paper and then hand it in to the teacher. It helps them to retain these concepts and, for us, to verify if what the majority of the course highlights is, in fact, what we intended them to retain; and, if it is not, to make us aware of the need to return to these topics in the next class. In general, it is advisable not to give it a name so that they do not feel that it is an exam or assessment.

Debates: To carry out this activity, the course is presented with two or more different positions on a given topic (if they are antagonistic, so much the better). The course is divided into two or more groups (as many as there are positions); and each of them is asked to spend a few minutes finding reasons to defend the position they have been assigned. It does not matter whether or not they agree with it, they have to find reasons to be able to defend it. After a reasonable amount of time for them to complete the task, a debate is generated between the different groups, moderated by the teacher. After a while of debate, the teacher takes advantage of the different opinions presented to channel them towards the topic that the students are supposed to learn. This type of activity not only encourages students to become actively involved in their learning, but also to learn to debate different ideas with consideration and respect for the opinion of others. We really need this skill nowadays!

Dramatizations: this activity can be prepared from one class to the next. A group of students (or several) are asked to research and develop a topic through role play. This role play can be a simulation of a professional situation related to the topic they are expected to learn about. It could be presenting a business proposal or a work plan, discussing a particular case with a client, resolving a problematic situation in the workplace, etc. In these cases, those who are not participating in the role-play should be assigned a role or, failing that, they should evaluate the role-play in order to make sure they are paying attention to what their classmates are doing. The teacher should pay attention to see how they can channel the activity in a way that ensures that the students achieve the desired learning outcomes.

Flipped learning (FL): this methodology basically tries to make the most of the time spent in class for interaction between teacher and students, and between the students themselves, leaving passive activities for a time prior to the face-to-face encounter. Although there is previous history, this methodology became popular at the beginning of this century thanks to the work of two North American educators, Jonathan Bergmann and Aaron Sams. An appropriate definition of the method could be the following:

> FL is a pedagogical approach in which first contact with new concepts moves from the group learning space to the individual learning space in the form of a structured activity, and the resulting group space is transformed into a dynamic, interactive learning environment where the educator guides students as they apply concepts and engage creatively involved in the subject matter. (Talbert 2017, p. 20).

There are those who limit the preceding activities to video viewing, but in reality, the concept is much broader, as is well expressed in the above.

Finally, there are many websites on the Internet that describe these and other strategies for promoting active learning. One of my favourites is the one developed at Harvard University under the name "Project Zero" (Harvard Graduate School of Education). There, activities can be searched according to discipline, type of competition and project; many of them are also in Spanish. Another good reference, although it is only in English, is the guide published by the Professional Development Service for Teachers of the Irish Department of Education (PDST). Lastly, one that is in Spanish is developed by the Center for Teacher Development of the Pontifical Catholic University of Chile (PUC de Chile).

2.2 Authentic Learning

In simple terms, we could define it as "learning by doing". Obviously, it also falls into the recently described category of active learning. As the renowned American pedagogue argues: "Learning to be a scientist is not the same as 'learning science'" (Bruner 1997, p. 232). Isolated facts and formulas do not acquire meaning or relevance until students discover their meaning and application within the framework of a specific learning activity. It is, therefore, about learning by solving complex real-life problems. By saying "complex", I do not mean "difficult", but rather that they involve different aspects or different disciplines, because real problems are rarely monodisciplinary.

To clarify, I will give an example. The principles of electromagnetism can be taught by presenting the equations of Maxwell's laws and asking students to solve theoretical exercises on paper (an activity that many mistakenly call "practical work"). Or very simple concrete experiments can be carried out with cables, batteries, magnets, electromagnets and other easily acquired objects; and through the development of these activities, students can discover and construct knowledge that will later be formalized through relevant explanations, a strategy also known as "inductive learning".

I suggest watching the video in which MIT graduates are asked to solve a simple practical problem:

https://youtu.be/aIhk9eKOLzQ

To clearly illustrate what I am referring to, I want to share something that I just heard in the staff room at my school. A math professor said: "I got tired of telling them that continuity is analysed in the domain, but they keep analysing it at a point that does not belong to the domain." For someone who has learned mathematical analysis, this makes a lot of sense, but for first-year students, it probably means nothing; it is one of those phrases that, in the best-case scenario, they repeat without understanding or apply mechanically. The problem, in my opinion, is not that the students do not know it and, therefore, do not apply it correctly, but that the teacher has not been able to contextualize that knowledge within the framework of something that makes sense to them, what Ausubel defines as "meaningful learning". (Ausubel 1960) (a concept to which I will return later in this chapter).

Another way of approaching this type of learning is to present students with real-life problems (selected by teachers with a specific intention and objective) and ask them to try to imagine possible solutions to these problems. It will be essential to provide them with the necessary scaffolding (we could call it the "clues"), to guide them on this path of discovery and construction of knowledge.

The characteristics of authentic activities are as follows (Oliver y otros 2002):

- Be relevant to the real world.
- Be ill-defined, requiring students to define the tasks and subtasks necessary to complete the activity.
- Include complex tasks in which students must research over a sustained period of time.
- Give students the opportunity to examine the task from different perspectives, using a variety of resources.
- Provide the opportunity to collaborate.
- Provide the opportunity to reflect.
- Be able to integrate and apply in different subject areas, and lead beyond the specific outcomes of a domain.
- Integrate seamlessly with assessment (we will develop this topic later, when we talk about assessment).
- Create polished and valuable products in their own right, and not as preparation for something else.
- Allow for competitive solutions and a diversity of results.

These types of authentic learning strategies can be combined with flipped learning in a very effective way.

It can be argued that this is not always possible because physical or economic resources do not allow for the implementation of these types of strategies. And that is when technology can be an important ally in the form of simulations, remote laboratories, extended realities and others that I will refer to later in this book.

2.3 Collaborative/Cooperative/Team Learning

While in pedagogical terms there are those who differentiate between collaborative and cooperative, personally, I do not think it is relevant to clarify this here and now, given the purpose of this text. Indeed, in the Dictionary, the meanings of cooperate and collaborate are similar; in fact, it indicates that they are synonyms. What we are referring to here, to be more specific, is teamwork in which the following conditions must be met (Felder and Brent 2016):

1. Positive interdependence.
2. Individual accountability.
3. Promotive interaction.
4. Development an appropriate use of teamwork skills.
5. Regular self-assessment of team functioning.

One of the issues that is important to emphasize is that this type of strategy not only enables and facilitates the learning of concepts and skills specific to the discipline in question, but also fosters the development of skills in teamwork, conflict resolution, role taking, shared time management, etc.

In the previously mentioned book by Felder and Brent (2016), you can find a series of suggestions and recommendations for organizing this type of learning strategy. For those who cannot access this book, I will share some practical advice that comes from reading the aforementioned book and others no less valuable, as well as from my own teaching experience:

1. **The make-up of the team:** The optimum number is three or four members and, as far as possible, they should be of a mixed nature. When groups are formed according to the affinities or friendships of the students, these groups tend to perform much better than the others, and this is not beneficial for all the members of the course. Furthermore, as I usually tell them, in real working life, groups are not formed by friendships or affinities, so it is very necessary to get used to working with people you do not know.
2. **Learning to work in a team:** Firstly, it is important to emphasize that a team must be formed; it is not a mere grouping of people, but a group of people who collaborate to achieve a common goal in which each of its members must play a specific role. Now, teamwork is not an innate skill; on the contrary, it is something that students have to learn; it is our responsibility as teachers to guide them so that they achieve it.
3. **Allocation of roles and establishment of responsibilities:** As part of what I mentioned in the previous point, it is important for team members to define the roles they will play and what responsibilities each of them has. It is also important that these roles rotate throughout the school year for each of the projects that will be undertaken, whenever possible.
4. **Work plan and meetings:** It is also important for the group to establish a work plan with dates and times that allow for reasonable and effective compliance with

the deadlines proposed by the teacher. Likewise, the times of the team meetings and with the teacher.

5. **Submission or presentation of the work:** It is advisable, whenever possible, to schedule partial submissions in order to be able to monitor the progress of the work and avoid delays that might jeopardize its timely completion. It is also advisable to define the form in which the work will be prepared and presented, and (as mentioned in Point 3) what the responsibilities of each member will be (in the case of an oral presentation in front of the class).
6. **Work tools:** Nowadays, work teams make intensive use of digital collaboration tools, such as videoconferencing, messaging, shared documents, etc. In this sense, it will be convenient for the teacher to decide or agree with the course which tools will be used in order to avoid dispersion that undermines the teacher's ability to effectively monitor the progress of the work.
7. **Assessment:** Although I will deal with the subject of assessment in detail in the corresponding chapter, it is worth mentioning here that it is very advisable to include peer assessment in this type of activity, in addition to the necessary teacher assessment.

The strategy of "peer learning" is worth mentioning here. But before moving on to a description of this strategy, I want to point out something that is explicitly or implicitly mentioned in several parts of this book: we have to banish the idea that the teacher is the only one in the classroom who "knows" and, therefore, teaches; and that the students "do not know" and, therefore, are the only ones who learn. In a modern classroom (by which I do not exactly mean issues of infrastructure or technology), the aim should be to promote an environment that we could call "enriched learning", in which everyone (teachers and students) has something to contribute and to receive. Experienced teachers know that we continue to learn when we teach; and, consequently, it is very beneficial for students to teach in order to learn. It is not a question of delegating our responsibility to them; on the contrary, it is a question of them valuing and reinterpreting what they learn by helping their classmates to understand the concepts that are most difficult for them. On the other hand, when a student explains something to a classmate that they have not yet understood, they surely do it in a different way to us, from a different perspective, one that is closer to them. This idea is related to the assumption (a very mistaken one, by the way) that people (and our students, in particular) all learn at the same pace. Consequently, there will be those who learn faster; and, instead of "rushing" our classes to their pace, thus leaving behind those who need more time to understand, it would be more effective to allow and encourage this strategy that re-signifies roles in the classroom while promoting collective and more even learning.

One of the most active promoters of this strategy is Eric Mazur, who developed at Harvard University what he called "peer instruction", which consists precisely of involving students during class through activities that require each student to apply the basic concepts presented and then explain them to their classmates (Mazur 1997; Mazur and Crouch 2001).

As a final comment/suggestion, given how familiar young people are with the performance of sports teams, whatever the sport, it is advisable to use examples from that field to make the challenges and advantages of teamwork more understandable. And remember that we are the coaches of these teams, the teachers.

I also suggest watching this short video of a successful sports coach:

https://bit.ly/video_entrenador

2.4 Problem/Project-Based Learning (PBL)

There is no significant difference between the two strategies mentioned in this section, since the execution of a project involves one or more problems to be solved. It could be said that it is, in general terms, problem-based learning, organized in the form of a project. The following definition provides a little more detail on this:

> A project is a complex effort that necessitates an analysis of the target (problem analysis) and that must be planned and managed, because of desired changes that are to be carried out in people's surroundings, organization, knowledge, and attitude to life; it involves a new, not previously solved task or problem; it requires resources across traditional organizations and knowledge; it must be completed at a point in time determined in advance. (Algreen-Ussing and Fruensgaard 1990)

It is a strategy that, in some way, encompasses the three that I have previously described, as it involves active and collaborative learning and, ideally, is based on real-life problems or situations, in other words, authentic. Figure 2.2 shows how PBL is organized (Kolmos y otros 2004).

I think it is important to emphasize that PBL should be a self-directed and self-regulated activity for the students themselves, and that the teacher should only act as a tutor or guide for the learning process. It also promotes the development of professional skills, as it simulates the usual way of working in an organization.

2.5 Personalized Learning

I often say that the idea of one size fits all is very practical for certain items of clothing, but not for education. As I have said before, the education system that was created at the time of the Industrial Revolution, which aimed to train as many people as possible in the shortest possible time and which, consequently, treated students as cogs in a production line, is completely inadequate for today's society. Nor am I proposing that we go to the other extreme, that is, one-to-one education, as this

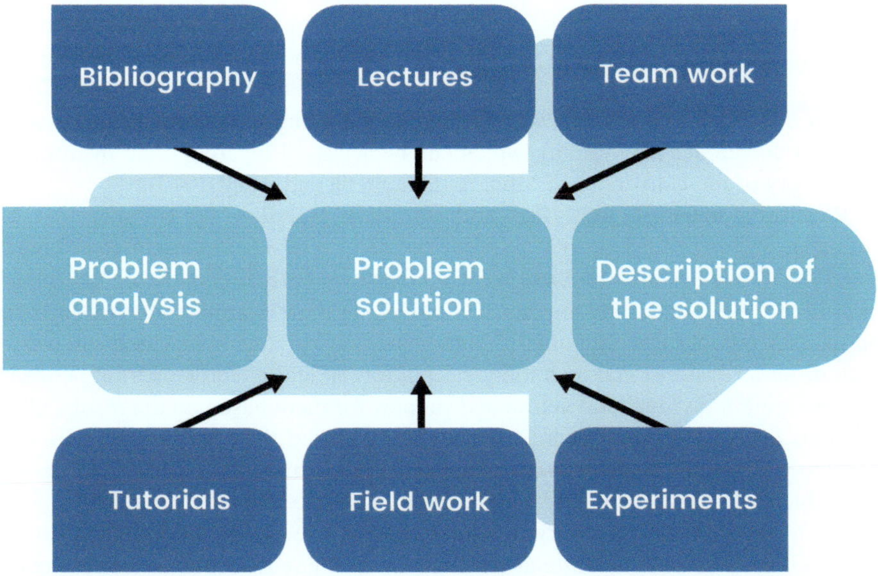

Fig. 2.2 Problem-based learning

is incompatible with the effective need for as many people as possible to complete their secondary education and, if possible, university education as well. However, there are strategies for achieving effective personalization of learning. Some can be promoted through institutional regulations; others, through activities proposed by teachers; and there are also those that are facilitated by technology.

Among the former is the aforementioned possibility of creating flexible learning pathways in which students can choose what they want to learn. One way of doing this is through elective subjects, although, unfortunately, in many cases, current curricula offer very few subjects of this kind. Another way is to provide intermediate degrees or specializations, as with the so-called majors and minors, widely used in North American universities, whereby the student first chooses a general area of study (such as engineering, social sciences, economics, etc.) and then a more specific specialization (for example, electronic engineering or macroeconomics).

As for the second possibility (that is, those associated with teaching), many of the ideas and possibilities that I have described in the different sections of this chapter are undoubtedly very effective in achieving the desired objective.

Finally, in terms of technology, I will mention two of them here: **personal learning environments and learning analytics and adaptive learning**.

Personal learning environments, also known as PLEs, can be defined as "a collection of tools established for the purpose of being used by a user according to their needs, fundamentally aimed at incorporation for personal work and, of course, for the development of learning activities". (Cabero Almenara and Llorente Cejudo 2015). Therefore, it is not a specific tool, such as a virtual learning environment

(VLE), also known by its acronym (LMS), but a set of tools that allow the student to personalize their learning experience. For those who are more interested in this subject, I suggest reading the book *Entornos Personales de Aprendizaje: Claves para el Ecosistema Educativo en Red* (Personal Learning Environments: Keys to the Online Educational Ecosystem), which can be accessed for free in the Institutional Repository of the University of Murcia. (Castañeda and Adell 2013).

Regarding **learning analytics and adaptive learning**: in a paper that my colleague and friend Eduardo Vendrell Vidal and I wrote (Cukierman and Vendrell Vidal 2020), we describe in detail these concepts that are directly related to the use of artificial intelligence and big data techniques. In short, learning analytics is based on the analysis of large amounts of data generated by students in their interaction with different institutional technological platforms; and adaptive learning consists of generating individual and automated support for them based on the results of the previous analysis. In this regard, it should be clarified that these techniques were developed long before the advent of large language models (LLMs) and derived tools, such as ChatGPT (a topic I will expand on in the respective chapter).

I would like to make an observation here that I believe is very important with regard to this issue. The use of artificial intelligence techniques to obtain results related to students and their academic careers raises issues that go beyond the academic framework and that delve into sensitive areas related, on the one hand, to data privacy and, on the other hand, to how algorithms process that data to obtain results that can affect our perception or opinion about another person. As stated in an article published by UNESCO:

> Algorithm-based learning raises new dilemmas in terms of justice. Power will be in the hands of those who control educational data. With enough data, an outcome can be anticipated, a trajectory predicted or modified. Those who control the platforms will be able to guide people's educational destiny. (Rivas 2021).

2.6 Deep Learning

I place this topic here because I understand that it is a concept that encompasses everything expressed so far in this chapter.

In the 1960s, David Ausubel introduced the concept of "meaningful learning", which is what happens when students make connections between new information and their prior knowledge, making it more likely that they will delve deeper into the subject. (Ausubel y otros 1983). Subsequently, several authors, including John Biggs, worked on another concept related to the previous one, known as "deep learning". "The deep approach stems from the perceived need to approach the task in an appropriate and meaningful way, so that the student tries to use the most appropriate cognitive activities to develop it" (Biggs 2005, p. 35). Closer in time and space, Melina Furman quotes David Perkins when she proposes the following definition of this type of learning: "understanding implies acting with knowledge in a *flexible* way". Furman also points out that this type of learning generates "transfer", understood as "the ability to use what we know in a different context" (Furman 2021, p. 53). Both meaningful and deep learning contribute to a more solid and lasting understanding of a subject. Meaningful learning provides a solid foundation, while deep learning allows students to delve into the complexities of the subject.

2.7 Student-Centered Learning

All the strategies described in this chapter have a common denominator, which I have already mentioned: student-centered learning (SCL). Education systems, and especially universities, are based on a model in which the most important person in the classroom is the teacher: the one who defines the rules of the game, the one who knows everything, the one who never makes mistakes (three conceptually and factually false assumptions). I return here to the question I asked at the beginning of this chapter. If our students do not learn, what's the point of what we teach them? That is why it is important to change the way we define and organize learning activities. Usually, we start by thinking about what we are going to teach (i.e., focused on the teacher) and we should start by thinking about what we want the students to learn (i.e., focused on the students). Typically, the sequence is as follows:

1. We define the contents or topics to be taught, typically many, because the course syllabus prescribes it (as if we were not the ones who define the syllabus) or because the author of the course book decided it that way.
2. We define the activities that we will carry out, generally master classes; and even when so-called "practical work" is included, it tends to be mechanical and repetitive problem-solving exercises.
3. We define evaluations, generally summative and based on the repetition of what the teacher explains (see Fig. 2.3).

Clearly, this sequence does not promote a SCL. An alternative proposal is presented in the book *Understanding by Design* (Wiggins and McTighe 2005), which could already be considered an essential classic for teacher training and which proposes an alternative sequence to what they call "inverse design":

1. Identify what students should know, understand and be able to do. What content is worth understanding. What lasting learning is desired.
2. Determine how we will know if the students have achieved the learning outcomes defined in the previous step. What will we accept as evidence of the students' understanding and competence?
3. With the two previous aspects in mind, we should design the most appropriate activities to achieve the expected learning outcomes (see Fig. 2.4).

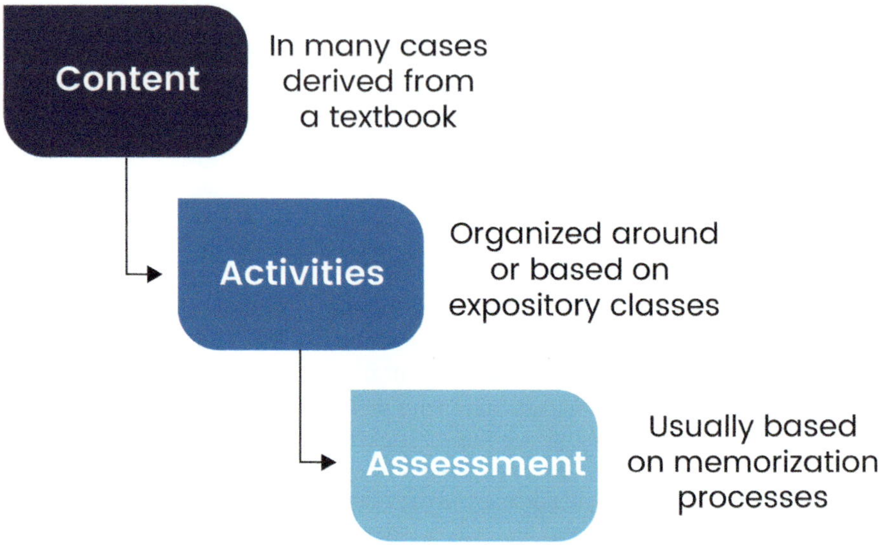

Fig. 2.3 Traditional sequence

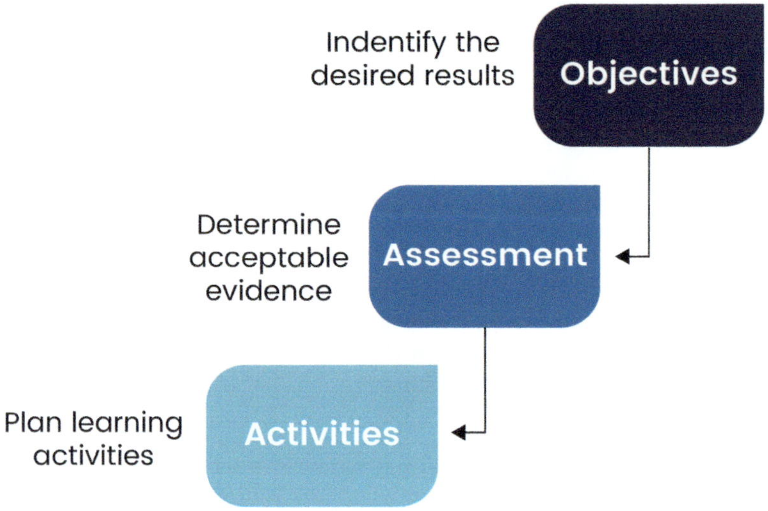

Fig. 2.4 Backward design

A very good summary of this idea, explained by one of the authors of the book, can be seen in these two videos:

https://bit.ly/UBD_Parte1
https://bit.ly/UBD_Parte2

Biggs expresses a similar sentiment when he proposes what he calls "constructive alignment":

> A good teaching system aligns the teaching method and assessment with the learning activities established in the objectives, so that all aspects of this system agree to support the proper learning of the student (Biggs 2005, p. 29).

If we do not put the student at the centre of our concerns and our decisions as teachers, everything else we do, no matter how well-intentioned, will not obtain the best result. I would like to quote the words of the former Director of the UNESCO International Institute for Higher Education in Latin America and the Caribbean, Dr. Francesc Pedró, who, in a recently published article, referring to higher education, said:

> It is important that higher education does not lose sight of the fact that one of its missions is to attend to the needs of people who see in it a hope for personal and social development. Placing people at the centre of its educational mission is the best bet for transformation that it can make (Pedró 2023).

2.8 Learning to Learn

I usually say in my courses for teachers that much of what we teach today will be obsolete by the time our students graduate and that much of what they will need to know from that moment on does not yet exist. This assertion is endorsed by a report from the European Commission which states that students entering primary education today will probably work in jobs that do not yet exist, use technologies that have yet to be invented or face unexpected global challenges (Sala y otros 2020). This brings us back, as teachers, to the need to focus on the concepts that will not change and on the skills necessary to continue learning independently. In other words, learning the fundamentals of the discipline in question is as important as acquiring the ability to continue learning throughout life.

As expressed in the UNESCO report of the International Commission on Education for the twenty-first Century, the concept of lifelong learning is one of the keys to the century and responds to the challenges posed by a rapidly changing world, concluding that the only way to meet this challenge is for each individual to learn how to learn (Delors 1996). In the same vein, the aforementioned report (Sala y otros 2020) states that learning to learn is the most important of all skills.

Although the idea of "learning to learn" has gained some popularity recently, there are precedents on the subject going back several decades with, in some cases, different terms such as "strategic learning" or "self-regulated learning" (Gargallo López y otros 2020). We could talk, in general terms, about a student who can learn on their own: the idea is that they develop this ability to learn autonomously during their time in the classroom and then be able to deploy it throughout their life in the areas in which they work. Autonomous learners select and use self-regulated learning strategies to achieve the desired academic results based on feedback on learning effectiveness and ability. According to the aforementioned work, learning to learn competence involves five dimensions (see Fig. 2.5):

Now, the question you may be asking yourself here is how we develop this skill in our students. Certainly, it will not be through lectures and exams in which students' ability to repeat what the teacher has said is assessed. Quite the contrary, it is about using many of the strategies mentioned earlier in this chapter to promote the development of this competence. (Felder and Brent 2016) offer us a series of concrete suggestions to achieve this (see Fig. 2.6).

An example of inductive methods is the previously mentioned PBL. I will develop the topic of critical thinking later, in a specific chapter dedicated to the subject. Teaching metacognition involves helping people become aware of their own thought processes and guiding them to actively monitor, plan and evaluate their learning strategies. The latter is strongly related to the development of critical thinking. Finally, the challenges we present and the corresponding support we offer students must be in line with their level of education and maturity. There is nothing more demotivating than proposing activities that are far beyond the students' reach or, on the contrary, offering trivial solutions.

On the other hand, and as always happens in education, responsibilities are shared between teachers and students. Therefore, it is important to make explicit what they should do and how. Again, we turn to Felder and Silverman (1988, p. 235) in search of concrete clues, described in the form of questions that students should ask themselves (see Fig. 2.7).

Take a break from reading and watch a video.

The four pillars of learning:

https://bit.ly/cuatro_pilares

Perhaps, at this point, you are thinking that, if this is feasible to be carried out in a regular university course, why not in yours. In this regard, I can tell you about our own experience in the Communications and Networks course that we have been developing for 14 years in the Information Systems Engineering degree program at the Buenos Aires Regional Faculty of the National Technological University. When we started to introduce these changes, my colleague and friend Juan María Palmieri and I did not have much of an idea of how to do it, but we knew we did not want to keep doing what we had always done and it was not making us very satisfied, even though the results obtained up to that point were not bad at all. We learned how we wanted

Fig. 2.5 Dimensions of learning to learn

our students to learn, that is, **by doing**. You may wonder if we made mistakes; and the answer is that we made many, because there is no innovation without mistakes. I often say (although it is not very original) that the worst thing about mistakes is not making them, but repeating them, not learning from them. We have been brought up in a culture that penalizes error. That is precisely the most important mistake we have made, so it would be important to try not to perpetuate it in future generations. "Recognizing a mistake and taking advantage of it is a boast that borders on genius"

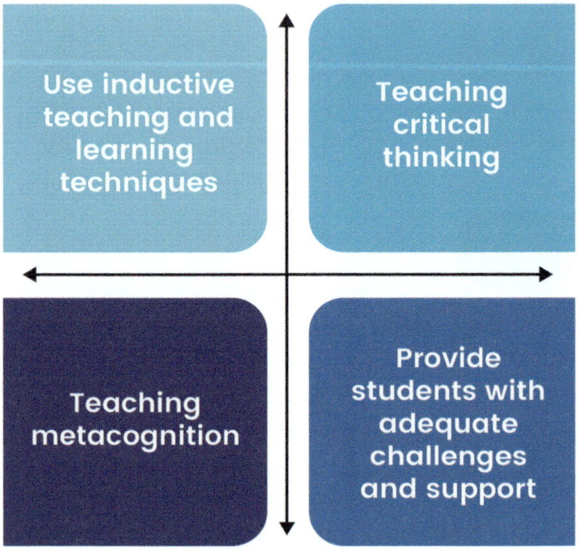

Fig. 2.6 Strategies for developing the competence to learn to learn

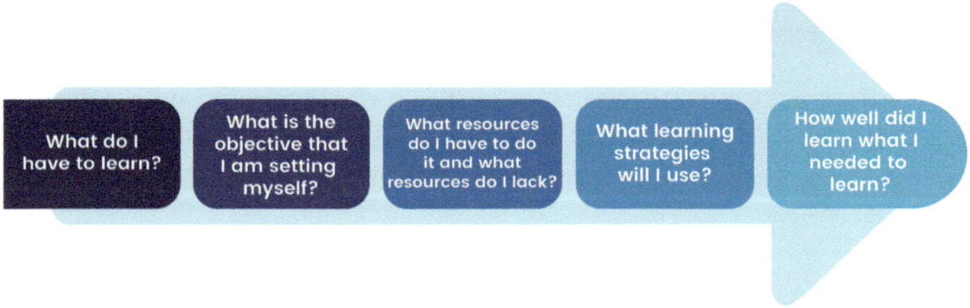

Fig. 2.7 Questions that students should ask themselves

(Marina 2006, p. 24). A renowned French cognitive neuroscientist says that "mistakes play a crucial role in learning. Making mistakes is the most natural way to learn; thus, learning and error become almost synonymous, because each mistake offers an opportunity" (Dehaene 2019, p. 258).

But, returning to our experience, the first thing we decided was not to try to change everything "all at once", but to introduce gradual changes, test them, evaluate them and correct and improve them with time and experience. To do this, we rely on the invaluable and indispensable opinion of our own students. Our proposal requires much more work and personal involvement than what they were used to; that is, coming to class (not always), attending class (not always), doing practical work (not always) and, finally, trying to learn what is needed to pass the midterms a few days (or a few hours) before the exam. At the beginning of the course, we explain

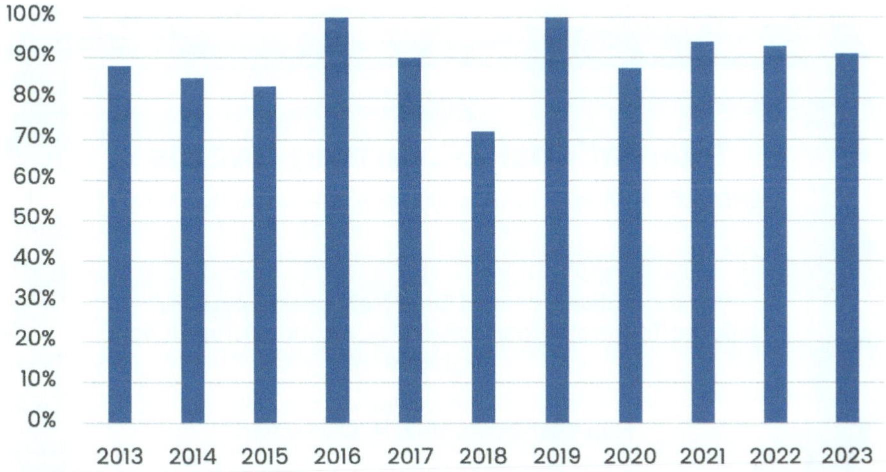

Fig. 2.8 Percentage of students who responded that they would choose the new methodology

the working methodology for the year; and, depending on the greater commitment and dedication required, the first reaction is usually annoyance (we could call it "resistance to change"). However, at the end of the year and despite the greater demands placed on them, the results were always excellent. Proof of this is the response to one of the key questions we ask in the anonymous final survey: "If you were given the opportunity to choose to study this subject with the traditional methodology or with the one we used during this year, which would you choose?" The responses were always, since we started with this modality, mostly positive (see Fig. 2.8).

I will reproduce just a few comments from last year's survey that clearly exemplify the opinion of the majority of students:

- Experience is the best learning method.
- I found this methodology more dynamic and bearable.
- I found it more dynamic and it allowed me to stay on topic all year round.
- Because it allows you to keep the subject up to date and, when you are a stickler for rigor (like me), it prevents you from leaving everything to the last minute.
- I prefer this modality because it reverses the traditional modality, in which one ends up studying everything last minute and from memory. This methodology proposes constant learning, where knowledge is incorporated with practice and connection of all concepts.
- More dynamic and entertaining, less stressful. I learned more than by studying for an exam and forgetting everything the following week.

Although the recognition of our students was always the best reward for the effort involved in implementing this methodology, I am also proud to say that, in 2022, we received an international award for our initiative, granted by the International E-Learning Association (see Fig. 2.9).

Fig. 2.9 E-Learning Association Award

Although only my name appears on this diploma, as I submitted the application, it should be made clear that the award was given to both teachers on the course.

To conclude this section and the chapter, I would like to quote the Argentine philosopher and writer who, in his latest book, says: "A true teacher teaches how to learn. And learning how to learn is possibly the most fruitful legacy that can be given to those who are being educated, since it discourages them from clinging to a certainty that they want to be immovable". (Kovadloff 2023).

To finish, take a look at this short video.

I recommend watching a video in which the prestigious historian and writer Yuval Harari talks about this subject:

https://bit.ly/Yuval_N_Harari

Chapter 3
About Assessment

I am against exams. They only discourage the interest in studying.
Albert Einstein

I always remember a professor from my undergraduate days who, on the first day of class, would ask all his students for their student ID, mark them as having passed the course, and then tell them that anyone who truly wanted to learn could stay and complete the course properly. To be honest, I was often tempted to do the same in my own classes, but I never quite dared to take it that far. However, what I do every year on the first day of class is ask my students, anonymously, whether they enrolled in the course to pass or to learn. Usually, about two-thirds of the class respond that they are here just to pass. So, with a serious face, I tell them that those students can leave because they have chosen the wrong course. After a few moments of confusion, I clarify that it was just a joke and that they are, of course, all welcome to stay. However, I stress that they will only pass if they truly manage to learn. I also remind them that we, as their teachers, will do everything in our power to help them succeed, but they must put in their own interest, dedication, and effort to achieve it.

For as long as anyone can remember, formal education has revolved much more around passing than learning; it has revolved much more around the formal confirmation that certain levels of the minimum required knowledge have been reached, obtained or achieved than around the effective certainty that students have learned something more than the minimum necessary to pass the subjects. I know that this is a controversial statement to say the least, but unfortunately there is a lot of evidence to support it.

> One problem with the systems of assessment that use letters and grades is that they are usually light on description and heavy on comparison. Students are sometimes given grades without really knowing what they mean, and teachers sometimes give grades without being completely sure why. A second problem is that a single letter or number cannot convey the complexities of the process that it is meant to summarize. And some outcomes cannot be adequately expressed in this way at all. As the noted educator Elliot Eisner once put it, "Not everything important is measurable and not everything measurable is important." (Robinson and Aronica 2015)

U. R. Cukierman, *The Art of Teaching, the Magic of Learning: Towards Innovative Higher Education Systems*, Studies in Systems, Decision and Control 697,
https://doi.org/10.1007/978-3-032-35021-3_3

Interested in knowing since when this has been the case, I consulted the new digital oracle, ChatGPT, about the origins of assessment in the history of formal education. And it replied that in ancient China, from the end of the sixth century, what became known as the "imperial examination system" was established, which was used to assess candidates based on their knowledge of the Confucian classics and served as a means of selecting people for government posts (OpenAI 2024). Much closer in time, I remember the heated discussions that took place during the COVID-19 pandemic about how students were going to be assessed and how cheating in exams was going to be prevented. So much so that one of the most lucrative businesses at that time was so-called "Proctoring", i.e. systems that promised to prevent cheating in exams. One of these systems advertises on its website: "Our platform guarantees the integrity of the exam to protect the reputation of your institution and the success of the examinees." Now that we are in the age of artificial intelligence, quite a few systems have emerged, with their corresponding lucrative interest, which promise to detect when a text was written with this resource, following the same logic of identifying students who want to "cheat" in their assessments. This repeated situation reminds me of that old Tom and Jerry cartoon in which the cat is always chasing the mouse, trying unsuccessfully to catch it. Unfortunately, in many cases, when it comes to assessment, teachers become like Tom in the movie, and students like Jerry.

Another of the most frequent controversies, when discussing the education system, is standardized exams, such as those that are currently (January 2024) being tried to be made compulsory at the end of secondary school in Argentina. On this particular point, it is also appropriate for me to quote Sir Ken Robinson again, who, in one of his masterful talks, seen by more than 10 million people worldwide, said:

> The role of a teacher is to facilitate learning. That's it. And part of the problem is, I think, that the dominant culture of education has not been focused on teaching and learning, but testing. Testing is important. Standardized tests have a place. But they should not be the dominant culture of education. (Robinson 2013)

Personally, I believe that the conflict with assessment in education begins when we see it as something alien or separate from the learning process, and not as an integral part of that process. Having said that, you may be wondering why I am dealing with it in a chapter separate from the previous one. And my answer is because I believe that assessing is fundamental to learning and deserves all the dedication and analysis we can give it; and, perhaps, this way, we will understand that assessing is not the same as evaluating and that the use of exams as an instrument for assessment is not always necessary, much less exclusive.

Before continuing with the reading, you are invited to do an activity that will only take a few moments.

➡ Activity 3.1: What is the first response that comes to mind when you are asked the following question:

What is the first word that comes to mind when you think about assessment?

What did you answer? Most likely, you thought of an exam or any other type of traditional assessment. If so, do not worry, that is what most teachers answer. I have verified this in many courses where I ask the same question. If that is not your case, all the better. But please continue with this second activity.

➡ Activity 3.2: Food for thought

Do the assessment strategies you use help you to know if your students really learned what you wanted them to learn?

Do these strategies help your students to improve their learning? How?

Whatever your answer may be, hopefully, when you finish reading this chapter, you will be able to validate what you are already doing or incorporate some modifications that will help you improve your students' learning. But let's take it one step at a time. Let's start by analysing the terminology we use so that we can more easily gain a thorough understanding of the subject at hand.

3.1 Some Conceptual and Semantic Clarifications

Confusion often arises between the terms 'assessment' and 'evaluation'. In English, the term 'assessment' refers to ongoing measurement and feedback aimed at supporting learning and fostering improvement, while 'evaluation' is typically associated with making final judgments about the overall quality or success of a process or performance.

The assessment can be used to improve the quality of student learning and instructor teaching (formative), to assign grades or make pass/fail decisions (summative), or both (Felder and Brent 2016, p. 155).

I will transcribe the definitions that, along with colleagues from the Federal Council of Deans of Engineering (CONFEDI), all members of the Accreditation Commission, we developed when we were preparing the new accreditation standards for engineering degrees:

> **Assessment**: It is one or more training processes that serve to identify, collect and prepare data that make it possible to determine the achievement of learning outcomes. The assessment

can use both qualitative and quantitative methods, depending on the learning outcome to be verified, and should be understood as a process of improvement.

Evaluation: Consists of one or more processes to **analyse or judge the data and evidence accumulated through the assessment**. It is typically **summative** and involves the certification of a level or degree. (CONFEDI 2017)

Other important characteristics of the assessment are the following:

It should be known: Students should be informed in advance about how they will be assessed. In some cases, students could even be involved in defining the assessment strategies to be implemented; especially in advanced university courses.

It should provide feedback: Students should receive the information obtained through the assessment activities in a way that helps them to make the necessary adjustments to achieve the expected learning outcomes.

It should be continuous: This does not mean that it should be done every day, but with a frequency that allows students to adjust their learning process. On this point, I often refer to a GPS: it would be useless to have a GPS that tells us we have taken the wrong road when we are already far from our destination.

It must be individualized: This premise is sometimes difficult to achieve in very large courses, but in those cases, the appropriate technology can be a good ally. I will delve deeper into this topic later in the book.

It must be positive: Understanding it as such that it reinforces the process of improving learning. Positive does not mean that it is just to say that everything is fine, but rather that, even when there is learning that has not been satisfactorily achieved, the feedback provides information and resources that allow the student to know what they are doing well, and in what and how they should improve.

All these characteristics clearly point to an assessment of processes, rather than just results; and that it should be understood as part of the learning process. "Assessing should become a reflective process where the learner becomes aware of themselves and their goals, and the teacher becomes a guide who positions towards the achievement of cultural and educational objectives" (Bordas and Cabrera 2001).

Having made these points, let's move on to some questions that, perhaps, we all ask ourselves at some point in our training or teaching experience referring to assessment.

3.2 What Types of Assessments Are There?

Assessing **LEARNING** is the most traditional method. It is summative and, therefore, takes place at the end of a period; typically, the so-called "partial exams". Its intention is to determine if the students have achieved the expected learning in that period and it is usually used for the corresponding grade. When it is the only form of assessment, it has several drawbacks, namely:

- The end of a period (two-month, four-month or semester) could be too late for the student to have the first information on how they are progressing with their learning.
- Due to the time that has elapsed, it tends to include a lot of content, and it is more difficult to determine what has and has not been learned.
- It generates a level of stress in students that does not help them to effectively demonstrate their knowledge. In many cases, it is more a test of their capacity to face stressful situations than of what they have learned.
- Due to time or instrumental reasons, it is often assessed in a different way to how it was learned. For example, computer programming is learned by working in pairs and is assessed on paper individually.
- It ends up assessing the result more than the process.

Given all of the above, it is clear that this is not the most advisable type of assessment, unless it is a complement to other assessments, such as those described below. Perhaps, as Einstein said, exams "only discourage interest in studying" (Calaprice, The ultimate quotable Einstein 2011, p. 108).

Assessment **FOR** learning is formative and, therefore, should be carried out during the learning process. It serves students, like the GPS I mentioned earlier, to keep track of how their learning is progressing and to make the necessary corrections to reach their destination (learning) satisfactorily; and, as a consequence, to pass the units and courses. But it is not only useful for the students, it is very important for us, their teachers, because it allows us to make the necessary corrections as we go along so that they achieve the expected learning outcomes. What is the point of developing the whole program in the time planned if our students have not managed to learn? Often, the worst enemy of effective learning is the demand to cover all the topics included in the program in a time that hinders or even prevents students from achieving a successful learning process. Perhaps we need to critically review the plans and see which of the learning objectives are important and feasible, but that is a different matter. I will return to this topic in the next chapter.

Diagnostic assessments are also included in this category, which are often only understood to be for the beginning of a course, when, in reality, they are also very effective throughout the course with the aforementioned objective of allowing students and teachers to know if the expected learning is being achieved or if it is necessary to correct the course in some way.

Assessment **AS A FORM OF** learning is a process of developing and supporting students' metacognition. In this case, the students are their own assessors, either through self-assessment or by their peers. I will elaborate on this topic in the next section of this chapter.

You might like to watch this video by Professor Rebeca Anijovich:

https://bit.ly/Eval_RA

3.3 What to Assess?

It may seem like a simple question to answer, but, as it is so trivial, we often do not ask ourselves. Being clear about what we want to assess is the first step in designing an assessment process. And I use the word "process" because, just as learning is a process, assessing cannot be an isolated event. If we understand that assessing is part of the learning process, then it must accompany that process.

Having made this clarification, let's move on to the question in the title of this section. What do we want to assess? Do we want to assess isolated knowledge or, perhaps, the application of that knowledge to the resolution of a series of theoretical problems; or the application of a series of knowledge, skills and attitudes applied to the resolution of real-life problems or those of a specific discipline. The list of possibilities could be extended, but it is enough to understand that if we are not clear about what we are going to assess, we will be unable to design the assessment process. I am not making a value judgment here about what is going to be assessed, but rather emphasizing that, first of all, we must be very clear about what we are going to assess. In the previous chapter on teaching and learning, I described several different ways of learning. Consequently, the assessment should be adapted to the type of learning activities we have adopted. As I have said previously, in this matter, there is nothing worse than designing a learning activity that is then assessed without taking into account how the activity was carried out.

In the section on student-centered learning, I mentioned backward design as a methodology and constructive alignment as a concept. I quote it here again to save you having to look for those lines further back in the book:

> A good teaching system aligns the method and assessment of teaching with the learning activities established in the objectives, in such a way that all aspects of this system are in agreement to support the adequate learning of the student. (Biggs 2005, p. 29)

Returning to the backward design model proposed by Wiggins and McTighe (2005), and analysing it from the perspective of the alignment to which I refer, the diagram in Fig. 2.4 could be reframed as follows (see Fig. 3.12):

The reasoning that should be followed is as follows:

1. The first step is to decide what I want my students to learn. Moving forward without being clear about this is like going on a trip without knowing where I want to go. For example, I want them to learn how to solve differential equations per se (which would be very valid in a degree in Mathematics) or I want them to learn how to model a physical process using these equations (essential, for example, in an Engineering degree). I want them to learn what happened in the Second World War or to understand how that war changed the geopolitical logic that had prevailed until then. I could go on mentioning examples, but I want to make two points clear here. One is that all the examples I have described are valid, they are not mutually exclusive as long as I am clear about what I want my students to learn. The second issue is that I have not mentioned examples of the type I want them to know about the rivers of Europe or the countries of Latin

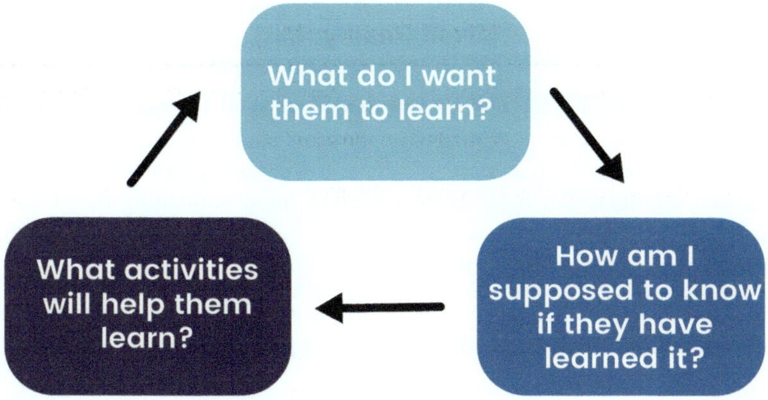

Fig. 3.12 Constructive alignment

America; not because I believe that this information is not valuable, but because it should be learned in context, not as a mere memorized statement.

I would like to quote here a paragraph from a recently published book that I am reading at the moment, while writing the book:

> Assessment gives us a measure of what and how much has been learned, but above all it implicitly communicates what needs to be practiced. It is the "value function" of a student's neural network. If a history exam asks for a list of dates of deaths and births, this is how they will structure their memory. If they are asked to retain information in a lasting, associative way, intertwined with the rest of their thinking, deep learning is promoted (Sigman and Bilinkis 2023, p. 111).

2. The second step is to define what evidence will **allow me to know if the students have achieved the expected learning outcomes**. Is it useful for me to verify that they know how to solve a series of theoretical exercises or do I need to know that they are capable of solving real problems? Does describing the facts with dates and protagonists allow me to know if they have learned the importance of the Second World War for what happened later in Europe and the world or do they need to analyse the political, social and economic effects in the post-war world? Surely, each of you can find other examples appropriate to your own discipline.
3. The third step is to **define the activities that students should carry out in order to achieve the expected learning outcomes**. Continuing with some simple examples related to the cases mentioned above, would it be advisable for them to carry out an infinite number of calculation exercises of increasing difficulty based on the methodology explained by the teacher, or would it be more effective for them to carry out an inductive process (I will come back to this method later)? Would it be useful for all the students on the course to read the same history text on the Second World War or to listen to the teacher's class? Or would it be better for several groups to read and analyse different texts, and then organize a class discussion coordinated and guided by the teacher?

4. Once we have answered the above questions, we must critically analyse whether what we wanted them to learn is possible to achieve in the time and conditions we are in; and, if so, what assessment strategies will be the most appropriate to verify if they have learned it; and then analyse if the planned activities are the most conducive to achieving the learning objectives, and so on, **until we are satisfied with the decisions made or until reality shows us that we have to make changes**. We must always bear in mind something that I have said previously: if students are not learning what we want them to learn, we have to change the way we are doing things.

In the following sections of this chapter, I will continue analysing different aspects of the assessment process that will help us to complement the analysis described so far.

3.4 Who Assesses?

What do you mean by, who assesses? The teacher, obviously. Who else could it be?

I have had this exchange with many colleagues in different fields but the truth is, the answer is not so obvious. In every classroom, there are students and at least one teacher. If we talk about assessing students (we could also talk about assessing teachers, which is necessary and convenient, but we are not going to deal with that now), there can be three types of assessment:

1. The teacher(s) assess the students.
2. The students assess each other.
3. Each student assesses him or herself.

In pedagogical literature, the first case is called "heteroevaluation"; the second, "co-assessment"; and the third, "self-assessment". But let's take one step at a time.

Heteroevaluation, as I have said, is what we teachers do and what I have been talking about since the beginning of the chapter, and to which I will also refer later on.

As for **co-assessment**, peer assessment is extremely important in two ways: for those who assess, because it promotes special attention to what their peers do; and for those who are assessed, because the opinion is from a peer who will most likely have a different view from the teacher. There is a widespread prejudice that students will assess each other positively, but in fact this does not usually happen, provided that the assessment strategy is well designed. It is worth remembering here that we are not talking about evaluating, but about assessing. Although, in some cases, a proportionally low component of peer assessment could be included in the teacher's assessment, so that it does not affect the approval or non-approval of what is being assessed. In these cases, it is very important that the criteria and the assessment instrument (I will talk about instruments later) are defined by the teacher or, in any case, together with the students.

Regarding **self-assessment**, an obvious definition would be to say that students assess themselves, which can occur in groups or individually. Going into a little more depth, it could be said that it is a process by which students, on the one hand, monitor and assess the quality of their thinking and behaviour when learning; and, on the other hand, identify strategies that improve their understanding and skills (McMillan and Hearn 2008). This concept of self-assessment is strongly related to the development of learning to learn skills, which I described in the previous chapter.

Rolheiser and Ross (2001) propose a virtuous circle model through which self-assessment contributes to learning (see Fig. 3.13). In this model, when students assess their performance positively, self-assessment encourages them to set higher goals (1) and to commit more resources or personal efforts (2). The combination of goals (1) and effort (2) equals achievement (3). A student's achievement results in a judgment about him or herself (4), such as when a student answers the question "Have I achieved my goals?" The result of self-judgment is self-reaction (5), that is, a student responds to the judgment with the question "How do I feel about it?" Goals, effort, achievements, self-judgment and personal reaction can combine to positively impact self-confidence (6). Self-assessment is, in reality, the combination of the components of the self-criticism and self-reaction model; and, if we can teach students to do it better, we can contribute to a virtuous cycle of better learning (Rolheiser and Ross 2001).

Finally, these same authors propose four stages to help teachers implement self-assessment in class, presented in schematic form in a work derived from the latter (see Fig. 3.14).

Much of what is described in this section can be adapted for application in the previous case of co-assessment.

3.5 How to Assess?

Answering this question has a lot to do with what we have already seen so far, but I will delve a little deeper into some specific issues. In the literature on the subject, there is no generalized agreement on what is meant by means, techniques and instruments of assessment. Therefore, abstracting from these classifications and trying to answer the question that begins this section in a clear and simple way, I will say that we have, on the one hand, the productions or performances of the students and, on the other hand, the tools that we use to evaluate and grade these productions. It should be clarified that the latter can be used, as appropriate or as decided, both by us, the teachers, and by the students when a peer assessment is carried out.

Let's start with student productions (or performances), which can be written, oral or practical, or a combination of these. Thus, we have the classics: written exams in all their variants, reports, essays, posters, concept maps, journals, monographs, podcasts, videos, presentations, debates, participation in forums, practical work, research work, field studies and all other student production that contributes to their learning and, at the same time, can be used as evidence of that learning. A special mention for

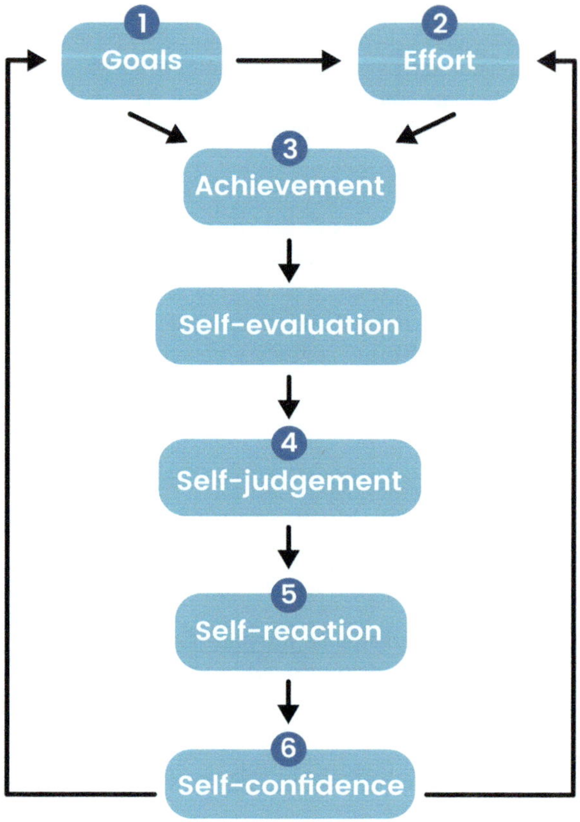

Fig. 3.13 The contribution of self-assessment to learning

portfolios, which are "a collection of work produced by students that reveal their progress over a period of time. They are based on reflection, which allows the learning process and the achievement of the proposed goals to be evaluated" (Anijovich and González 2021, p. 111). Nowadays, when we talk about student productions, it makes no sense to distinguish between those that are produced on paper or in any digital medium, obviously including the Internet.

Portfolios are extremely useful for assessing the process; as is commonly said, they allow you to "see the movie rather than just the photo". When used within the framework of an institutional strategy, they allow students to demonstrate the evolution of their performance across different subjects and at different levels; and they allow teachers to assess the student's work at a given moment in terms of the evolution that the student has made throughout their training or career. As I often say and repeat, what is important is the assessment of the student, rather than the level they are at, at a given moment; and this has a lot to do with the much-vaunted and little-practiced "inclusion". Let's take the case of a first-year course in any university degree. In all likelihood, the level of a student who studied at a good secondary school in the context of a family that gave them appropriate support will not be the same

	Stage 1 →	Stage 2 →	Stage 3 →	Stage 4
Level of Implementation	*Establishing Criteria*	*Teaching Students How to Apply Criteria*	*Providing Feedback to Students on Application of Criteria*	*Setting Learning Goals and Strategies*
Beginning	Criteria given to students for their reaction	Examples of applying criteria given to students	Teacher provides feedback	Goals and strategies determined by teacher
Intermediate	Students select criteria from a menu of possibilities	Teacher describes how to apply criteria	Feedback provided by both teacher and students	A menu of goals and strategies is provided by the teacher
Full	Students generate criteria	Teacher models how criteria apply	Teacher engages students in justifying their feedback	Student constructs goals and strategies

Fig. 3.14 Growth scheme for teacher implementation of stages of student self-assessment (McMillan and Hearn 2008)

as that of a classmate who did not have the same opportunities. The results achieved by the former will probably be better and achieved in a shorter period of time. But perhaps the effort made by the latter to try to reach the minimum level required is just as valuable and worthy of recognition, and it is very important for this to be recognized by teachers.

I can picture myself in a hypothetical conversation (or perhaps recalling a real one) that takes place during a course on assessment I teach, where I am interacting with a teacher attending the training. They might ask how I choose among all these options. My first response would be that every resource or tool should be selected based on the specific objective for which it is intended. I would then suggest they ask themselves a few guiding questions, such as:

- Does it contribute to the student's learning?
- Is it in line with what I want them to learn and with the type of learning activity (remember the concept of constructive alignment)?
- Will it allow me, as a teacher, to verify whether the expected learning outcomes have been achieved?
- Will it help the student to get an idea of their progress and where they need to improve?

Anyhow, there is no single answer. It all depends on the teacher, the subject, the students, the moment, etc. It may be that, even in the same subject, at a given time and with a given group, one thing works better than another. The important thing is to be encouraged to try; and trying implies the possibility of making mistakes; and making mistakes generates the need to recognize the error and correct it; obviously without compromising the students.

Let us now develop the subject of the tools we use to assess and evaluate these productions, which is just as or even more important than what we have seen previously. In fact, several issues depend on their correct selection and use, namely:

- That they promote the use of criteria known to the students in advance.
- That they are fair and transparent.
- That they give students information on what they are doing well and what they need to improve (good feedback).
- That they make our task easier.

There are several tools that, if used correctly, serve the purposes listed above, such as checklists, which are "a series of aspects, characteristics, qualities, observable actions about a process or procedure that are usually recorded in a double entry table"; rating scales, which "help to record the degree of development that a student achieves in relation to a process or product and also make it possible to record its frequency or the degree to which it has been achieved"; and assessment matrices or rubrics (Anijovich and González 2021, p. 43). I will focus a little more on the latter, since, in my experience, they are the most widely used.

A rubric is a matrix or table (see Fig. 3.15) in which three elements must be included: the criteria, the levels and the indicators. These elements are related to what is to be evaluated; in general, a given competency or an expected learning outcome.

The criteria are the aspects that will be assessed. The levels of execution can be defined quantitatively, qualitatively (for example: insufficient, basic, advanced, optimal; or in a more colloquial way: needs more work, you are doing well, you have achieved it, etc.) or mixed. Finally, the indicators should describe the quality of the criteria to be evaluated and give the student information about what they are doing well and what they should improve. As I have already said, students should know in advance how they are going to be assessed; and, if some of these tools are used, they

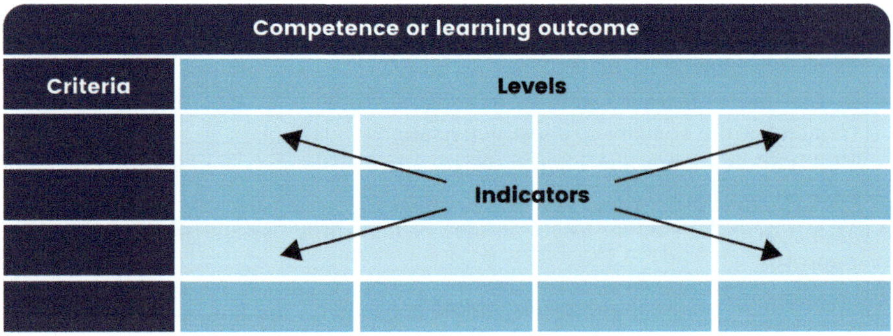

Fig. 3.15 Elements of a rubric

should be shared beforehand. Additionally, it can be a highly recommended practice to have students participate in the definition of these tools.

3.6 Authentic Assessment

In the previous chapter I described the concept of authentic learning. It is worth mentioning here what is known as "authentic assessment", a concept originally attributed to Wiggins, who states that it is "a true test" of performance or intellectual capacity because it requires students to demonstrate their deep understanding, their higher-order thinking and their ability to solve complex problems by carrying out exemplary tasks. Assessments should pose a real intellectual challenge for students, and teachers should design it in such a way as to achieve that goal (Wiggins 1989). This does not mean, as the author himself made clear in an interview, that any assessment that is not "authentic" is not good; in fact, he argues in favour of many of the assessment methods mentioned earlier in this text, such as portfolios, projects, etc. And he insists that no student should graduate without knowing how to do what is done in real life (George Lucas Educational Foundation 2002).

Meanwhile, authentic assessment is described as "allowing for the consideration of student heterogeneity and the possibility that all students will learn as long as they are offered a variety of activities in which it is possible to choose and make decisions to solve everyday problems" (Anijovich and González 2021, p. 15).

Returning to what has already been said in this chapter, the point is that assessments should be designed with the intention of forming part of the learning process and helping students to know what they are doing well and what and how they need to improve. Furthermore, they should help teachers keep a record of their students' learning; what corrections they should make so that the students achieve this learning; and, eventually, determine if they meet the requirements for approval and grade them.

Chapter 4
About Curriculum Design and Planning

A man who does not plan long ahead will find trouble at his door.
Confucius

Antoine de Saint-Exupéry is often credited with the following quote: "A goal without a plan is just a wish". Although the phrase does not appear verbatim in the famous book The Little Prince, the idea that a simple desire lacks the strategic steps necessary for its realization resonates in the topics found throughout the book. The little prince himself embarked on a journey in search of understanding and connection, a goal intertwined with a plan to explore the universe. Another book, very different from the previous one, but which has also had a great influence in various disciplinary fields is "The 7 Habits of Highly Effective People", in which the author explains that "To begin with the end in mind means to start with a clear understanding of your destination. It means to know where you're going so that you better understand where you are now and so that the steps you take are always in the right direction." (Covey 2003, p. 60). In short, it is clear that, in order to achieve a specific objective, it is necessary to have a concrete plan, to know what one wants to achieve and how one is going to do it. This idea can be applied both to the curricular design of an entire degree program and to the planning of a given subject.

Returning to the idea I expressed earlier that our objective as teachers should be for our students to learn, it is not only relevant but essential that we develop a plan to achieve this objective. And speaking of plans, we can mention those designed by the heads of departments and those designed by teachers within the framework of our own courses or subjects. Let's take one step at a time, from the general to the specific.

4.1 Curriculum Design

According to the Dictionary of the Spanish Language, curriculum means 'study plan' or 'set of studies and practices aimed at helping the student to fully develop their potential' (Real Academia Española). As Alicia Camilloni teaches us in a series of

U. R. Cukierman, *The Art of Teaching, the Magic of Learning: Towards Innovative Higher Education Systems*, Studies in Systems, Decision and Control 697,
https://doi.org/10.1007/978-3-032-35021-3_4

videos on the subject, there are different ways of understanding the curriculum; and, in this sense, she raises a series of questions for us:

- Is it what should be taught?
- Is it what is actually taught?
- Is it what the student should learn?
- Is it what the student actually learns?

Perhaps, the most appropriate answer, in my opinion, would be to say that the curriculum ought to be all of that, that is to say, there should be a coincidence between what should be taught and what is actually taught, and between what the student should learn and what they actually learn; but, in practice, that does not happen very often. They say that paper never refuses ink, but reality is often far from what is desired. That is why it is very important to understand, before moving forward with the subject, that the curriculum design should be a flexible and dynamic document, which is constantly adapted according to the results observed in its execution; and not, as happens in most cases, a text that remains unchanged for years or decades. All the more so when the curriculum in question corresponds to disciplines that evolve constantly and, at times, very rapidly, such as technology and, in general, the sciences. This does not mean that it is redone every year, but it does mean that the necessary corrections are made based on the results observed during its implementation.

But let's move on a little with the description of the components of the curriculum. Again, we are presented with a series of questions (see Fig. 4.1).

What do we want students to learn? Answering this question is directly related to the graduate profile. I will expand on this idea later.

How will we ensure that graduates successfully achieve this desired profile? To answer this question, it is first necessary to answer the questions I raised in the previous chapters.

Answering the question of **when** implies thinking about timing and circumstances. If we want the curriculum to be, as I said before, a flexible and dynamic document, a roadmap, the question of timing is not a minor issue. Internal and external circumstances condition the possibility of executing a particular curriculum. A paradigmatic case occurred during the COVID-19 pandemic: in practice, it was unfeasible to maintain the curriculum as it had been designed for normal times.

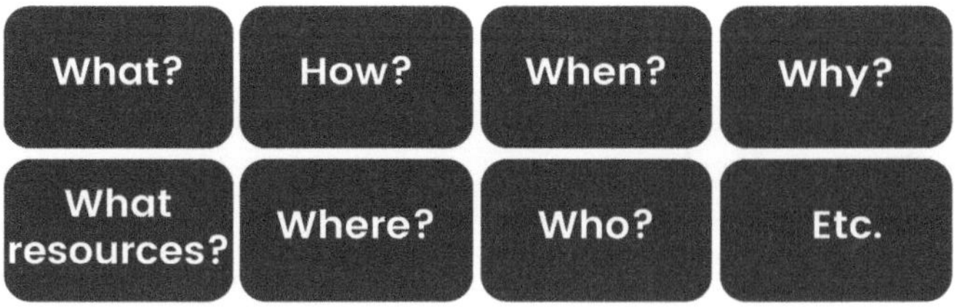

Fig. 4.1 Components of the curriculum

The question of **why** has to do with the objective we set ourselves when designing a curriculum and the possible answers to that question. Do we do it to comply with a formal requirement or to make it a living document that guides the development of a real training process? Although the answer we usually give is the second one, in many cases, we end up answering to the first one, although nobody admits it publicly. I will not mention names or specific references, for the sake of not naming names. I have witnessed this on many occasions and at different educational levels.

What resources will we use? This is not a minor question. Does the faculty in which the curriculum will be developed have enough and necessary resources in terms of buildings, furniture, physical, technological, administrative, etc., to satisfactorily carry out what is proposed in the curriculum? As I said before, paper never refuses ink. The important thing is to align the proposals with the real possibilities so as not to generate false expectations, which, in the end, conspire against the whole process.

Where? This question seems to have an obvious answer: the curriculum is developed in the faculty. But then I challenge that answer by asking another question. Only in the faculty? Implementing authentic learning (see Chap. 2) implies coming into contact with reality, with the outside world. Although there are already interesting experiences in this sense, they are not always taken into account when designing the curriculum; and that limits its possibility of practical implementation. After all, the professionals who graduate from our faculties will not work in the educational institution itself, unless they are going to dedicate themselves to teaching. For this reason, it is extremely important to include in the designs the guidelines to encourage the development of learning activities outside the classroom and in contact with professional reality.

Who? Another question that seems to have an obvious answer: teachers. And, again, I challenge that answer with a question. Is it true that only teachers can teach? What about the peer learning I mentioned earlier (see Chap. 2)? What about the possibility of bringing outside experts into the classroom to share their professional experiences? What about the possibility of creating learning communities with other institutions? These and other options should be considered in the design of the curriculum to enable their incorporation into everyday practices inside and outside our classrooms. Student-centered learning should start from the premise that teachers must lose centrality and that we have to make room for the inclusion of other actors when it comes to teaching.

Curriculum design is, then, the way in which the components of the curriculum are organized. Traditional curriculum design follows the logic presented in Fig. 4.2.

But this logic generally means including a lot of content because we assume that if we do not teach it, nobody else will; with the aggravating factor that this content is often out of date or will soon be. Then, the teaching methods tend to be very passive: theoretical lectures, practical work that becomes the development of a guide that leaves no room for true experimentation and discovery (and much less for error, an unavoidable element of real experiences), etc. Finally, formal assessment that, in many cases, require students to give answers or present solutions that have already been explained in class. This type of curriculum design responds to the following questions:

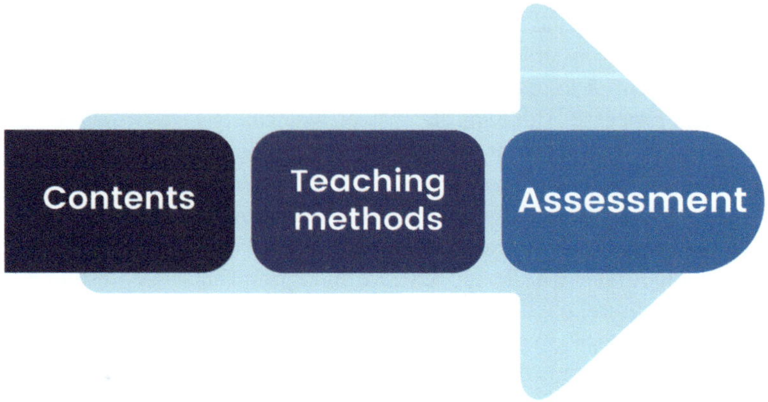

Fig. 4.2 Traditional curriculum design

- What content am I teaching?
- How can I cover this content during the course?
- What bibliography am I using?
- How am I teaching the class?
- How am I assessing?

I suggest you stop reading and answer a question.

➡ Activity 4.1: Who is the only protagonist in this story?

Have you answered it yet? Yes, you are right!! The only protagonist in this story is the teacher.

What if, when designing the curriculum, we changed the previous questions for the following ones?

- Who is it aimed at?
- What should students learn?
- How do they acquire knowledge?
- How do they develop skills and attitudes?
- How do they incorporate their personal qualities to achieve competencies?
- When is it certified that the student has achieved mastery of these competencies?

Who is the protagonist now? Yes, the student!! That is precisely what it is about. Designing the curriculum with the student in mind; and not with us, the teachers, in mind. It is also important to note that here I am not only talking about knowledge, but also about skills and attitudes; in short, about competencies, an issue that I have already developed previously and to which I will continue to refer in the following pages. In a competency-based curriculum design, the elements presented in Fig. 4.3 must be defined in an integrated manner.

Competency-based curriculum design is characterized by the following:

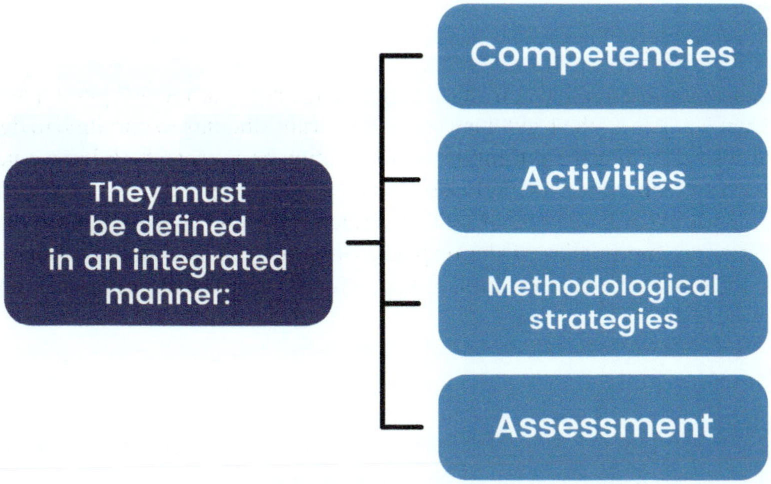

Fig. 4.3 Competency-based curriculum design

- Focus on performance, practice or application, and **not so much on content** (without believing that content is not important).
- Improve the **relevance** of what is learned.
- **Avoid the traditional fragmentation** of academic programs.
- Facilitate the **integration of content** applicable to work.
- Generate learning applicable to **complex situations.**
- Promote the **autonomy** of individuals.
- Transform the role of teachers towards a conception of **facilitator** (Yániz 2008).

Likewise, the following should also be considered:

- Identify and analyse the competencies that students should achieve (basic, generic and specific).
- Be organized by modules and problem-solving nodes.
- Encourage teamwork among teachers.
- Be based on "learning by doing", starting from the professional context and the integration of knowledge, attitudes and procedures.
- Have performance as the focus of the evaluation.

The phases of a curriculum design begin with the **analysis,** which arises from the adequate consideration of all the issues I have previously mentioned, which, in turn, will lead to the **design** itself. This process usually culminates in its **implementation**, and there is no phase for **assessing** what actually happens when the design is put into practice. Thus, as I have already mentioned, we find ourselves with curricular designs that were written on a desk, without much contact with the reality of what happens in the classroom; and which end up being, at best, a good declaration of intentions. Returning to the idea of a flexible and dynamic curriculum design, a permanent iterative process is required that adapts the curriculum design to the results of its

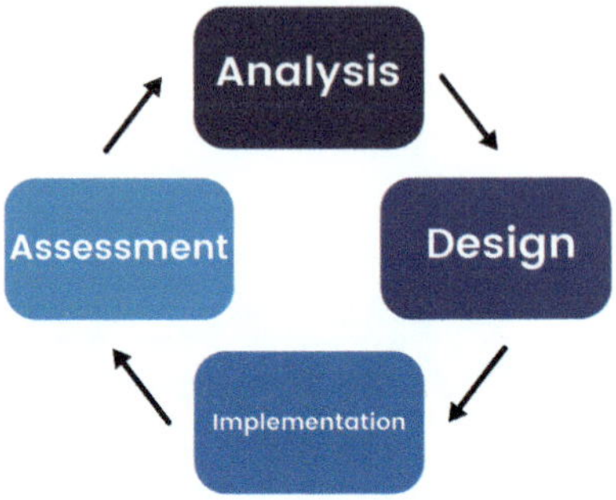

Fig. 4.4 Phases of an iterative curricular design

application, as well as to the changing conditions in which it is developed (see Fig. 4.4).

Going into a little more detail about the second stage of this process, the design itself, I propose an approach based on competencies:

1. Definition of the graduate profile
   Depending on the guidelines established by national regulations and accreditation standards (where they exist), and the institutional characteristics themselves.
2. Determination of the competencies that contribute to the development of this graduate profile
   These competencies can also be defined by higher authorities (such as those mentioned in the previous point), as well as by the regional or particular demands of each institution. Thus, the designs respond to the circumstances of the place where they are offered and of the institution itself.
3. Indicate the levels at which each competency and its indicators should be developed
   Three levels are usually defined for the development of competencies: low (L), medium (M) and high (H). Indicators must also be determined that will allow us to associate these levels with specific and concrete achievements to avoid, as far as possible, diffuse or biased interpretations.
4. Identify the subjects in which each competency will be developed at the required level
   This is the step in which the "curricular grid" is constructed, a matrix in which the subjects are located in the rows and the competencies in the columns, as seen in Fig. 4.5.

This grid could be understood as the "map" of the curriculum design (hence its importance). If the map has not been well designed, we will not reach the chosen destination. All the teaching departments of the degree course should be involved in this design, which should include processes of interaction and negotiation, in the best sense of the word, so that the different parties involved resolve the collisions or overlaps that this process inevitably generates. In short, it must be a team effort that guarantees not only its coherence, but also its feasibility.

The following are some important issues that need to be analysed in the grid:

	Generic competencies					Specific competencies				
Subjects	1	2	3	4	5	6	7	8	9	10
A		L		L			L		L	L
B	L				L					
C	L			L		L			L	L
D		M					M			
E			L					L	M	M
F	M	M			L		M			
G			L					L		
H	M		M			M		M		
I			H	M				H	M	M
J	H				H					
K	H		H			H		H		
L		H		H			H		H	H

Fig. 4.5 Curriculum grid

- Are all competences developed throughout the Curriculum?
- Is there a subject that should develop many competences or that does not develop any (review the rows)?
- Is there a competence that is not developed in any subject or is developed in too few subjects (review the columns)?
- Is the sequence of depth with which each competence is developed adequate?
- When assigning competences to subjects, was it taken into account that each one of them should develop, demonstrate and evaluate them?

It should always be borne in mind that competences already acquired and evaluated at the highest level in a subject or in a group of subjects do not need to be taught or evaluated again in other subjects, but they may be required by teachers and put into practice by students. For example, let's suppose that the skill of teamwork has been developed, demonstrated and assessed in subjects in the first three or four semesters of degree courses until a high level is reached. Then, in the following semesters, students may be required to work in teams, but it will be assumed that they already know how to do so. Conversely, we cannot expect students to work in teams if that skill has not been developed, demonstrated and evaluated beforehand.

5. Preparing the syllabuses for the subjects

   At this point, the syllabuses for the subjects should be designed in such a way as to develop the desired competencies at the required levels. I will expand on this idea in the next section of this chapter.

4.2 Course Planning

Teachers at all educational levels know that one of the first things we are asked to do when we take charge of a course (and sometimes, before we do so, when we apply to do so) is to develop the planning for that course. In my personal case, the first time I was asked to do this was when I faced my first course as a secondary school teacher at the age of just 21, with a lot of willpower but no preparation for the task. On that occasion, I was given a long spreadsheet known as the "A-B-C" (perhaps a colleague from that time remembers it), in which I had to fill in a series of fields that could be briefly described as follows: contents, activities, evaluation and bibliography. Today, more than 40 years later, many plans still have that format.

Once again, I suggest you stop reading and answer a question.

Activity 4.2: What is the first question you ask yourself when planning? What content do you have to cover? Or what do you want your students to learn?

Honestly, with one hand on your heart, what was your answer? If your answer was the first one, do not worry, we have been "formatted" to do it that way. But the good news is that there is still time to change it from now on. If your answer was the second one, congratulations! In any case, keep reading because you will find some additional ideas or resources to help you improve your teaching performance.

As I have already mentioned, there is usually a lot of content; perhaps taken from the textbook chosen for the course; following the logic of the author, which is surely not the same as ours, and in a context different to the one in which we will develop our course. Continuing with the components of planning, the activities in many cases are mostly passive and centered on what the teacher does; and the evaluation is fundamentally based on memorization or written or oral exams. If at this point in the reading you are thinking that what I am describing does not apply to you personally, congratulations once again. Fortunately, nowadays, many colleagues have already faced these changes.

Planning is an eminently human activity that represents "the expression of the desire of every educator to make their task an organized, scientific one, and through which they can anticipate events and foresee some results, including [...] the constant evaluation of that same process and instrument" (Pastorino et al. 1995). Planning is the practical and concrete interpretation of what is defined in the curriculum for our subject; and, therefore, it is our responsibility. It is the roadmap for the development of the course in our charge.

> Educational planning is simultaneously a **mental process** carried out by a teacher and a **product of that process**, a communicable, analyzable and modifiable product. This conceptual difference, which implies a necessary complementarity between the two, is important to highlight, **since any modification we want to make will have to define whether what we want to modify is the form that planning takes as a product, or whether we want to focus our best efforts on the way in which this mental, organizational, anticipatory and mediating process is carried out**. (Pastorino y otros 1995).

It should not be a mere formal document, but a working tool that serves the institution, the teacher and the student (see Fig. 4.6).

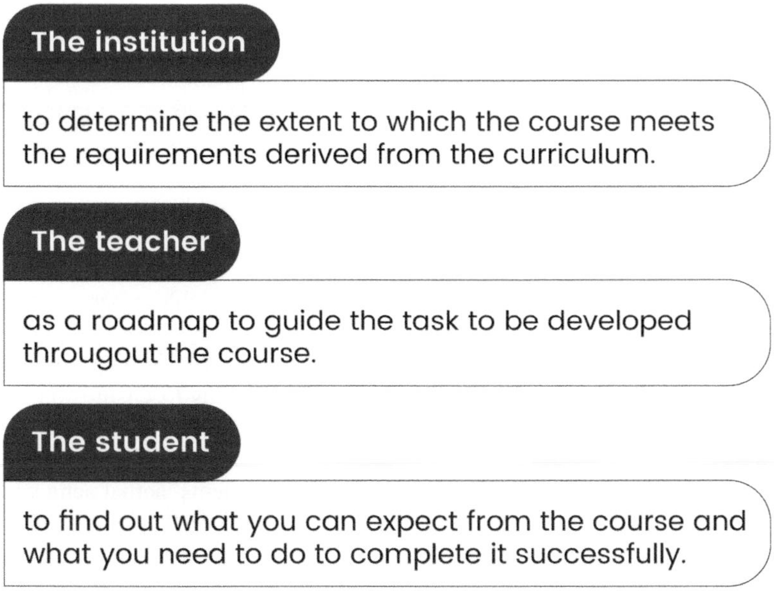

Fig. 4.6 Recipients of the planning

The planning must answer the following questions (Jerez y otros 2015):

- **What is learned?**
  - Content
  - Competences and learning outcomes
- **How is it learned?**
  - Strategies for development
  - Teaching and learning activities
  - Resources to use
- **How is learning demonstrated?**
  - Evaluation criteria
  - Evidence of learning
  - Procedures and instruments

Below are some important recommendations for planning:

1. Involve all members of the department in the development of the planning

   On many occasions, the planning is drawn up by the head of department without consulting or sharing the decisions with the teachers involved, which leads to a lack of commitment, if not a rejection of the guidelines expressed in the planning. Even when the head has to make decisions that are not shared by all the teachers, listening to them and explaining the reasons for the decisions leads to a better predisposition for the development of the course in accordance with the planning.

2. Collect and integrate students' assessments of the planning

   Students provide an alternative and complementary vision to our own, which, in general, is very valuable, as it shows us the reality that they perceive. In student-centered learning, this feedback should not be dispensed with. It can be obtained through informal conversations with students or through surveys, which should be anonymous to allow them to express themselves without limitations or fears that their opinions may affect their evaluations.

3. Make the organization of teaching activities more flexible in terms of learning achievement

   If the planned activities do not produce the expected learning outcomes, we will have to alter or add activities that will enable us to ensure that students acquire this knowledge or develop the skills that we consider essential in our subject.

4. Optimize the content to be covered based on the students' actual achievement of key learning outcomes. (**less is more**)

   The most important objective of our task is for students to learn what we consider to be essential in our courses. What is the point of "covering all the content" if students have not grasped the essential or fundamental learning points of the subject? Planning is a declaration of intent; and, as such, we must adapt it to the real circumstances in which the course is being developed, including the characteristics of each group, the particular moment we are going through, the limitations that may arise, etc.

5. Consider planning our "distinctive proposal" on how to teach a particular subject and get students to learn

   Although planning is common to all teachers in the department, each teacher and each group is different; therefore, planning should be adapted to these particular situations, thus acquiring the distinctive characteristic we are talking about. In this sense, it is important that the head of department understands and accepts the adaptations that, without altering the essentials, facilitate the achievement of the expected learning outcomes.

6. Being demanding and fair in the teaching and learning process does not prevent the generation of a pleasant learning environment.

   Demanding is often associated with maintaining a cold and distant relationship with students (nothing could be further from the truth). Good treatment and cordiality are essential for any relationship; even more so when the aim is to promote an atmosphere of trust and respect based on authority, which naturally produces a teacher who knows their subject and creates a pleasant environment, encouraging students to be willing to learn.

7. Test, adjust and update the planning periodically and frequently

   The plans are prior estimates of how the teaching and learning activities are expected to develop over the course. As I mentioned previously, reality may make us see that it is necessary to make adjustments and updates during the course of the course; especially at the end of one academic cycle and before starting the next.

8. Innovating inside the classroom

   Without innovation, there are no improvements. In other words, if you always do the same thing, you will get the same results. Often, the desire to innovate is limited by the fear of making mistakes, but "if you are not prepared to be wrong, you will never come up with anything original". (Robinson 2009, p. 38). The only bad thing about mistakes (as long as they do not lead to irreversible results) is repeating them. Therefore, try incorporating innovations that serve to improve learning with the peace of mind that, if you verify that they did not work, you can correct them without major consequences.

Having reached this point, I return to the approach proposed by Wiggins and McTighe (2005) with regard to "backward design" or "Understanding by Design", as the title of the book. In this sense, the authors propose a three-stage process, as described below.

Stage 1: Identify the expected results (what I want the students to learn)

- What should students know, understand and be able to do?
- What is it worthwhile for students to understand?
- What are the concepts that require lasting understanding?

Figure 4.7 shows the order of priority in which the topics should be analysed. Although there are limitations to this proposed order, given the necessary concatenation of the topics (for example, it is not possible to look at the concept of derivatives in mathematics before looking at the concept of limits), a critical analysis of the order of the topics in the planning is necessary, trying not to leave the most important ones to the end, as it is to be expected that, at the end of the course, we will not have enough time to develop them in the best way.

To determine these priorities, we have to ask ourselves to what extent the idea, the issue or process…

- represents a "big idea" that has lasting value beyond the classroom.
- is central to the discipline.
- needs to be discovered/interpreted.
- offers the potential to involve students.

Stage 2: Determine the acceptable evidence

At this stage of the process, we should ask ourselves the following questions:

- How will we know if the students have achieved the desired results and met the minimum required standards?
- What will we accept as evidence of a student's understanding and competence?

Stage 3: Planning learning activities

In this last stage, we should ask ourselves the following questions:

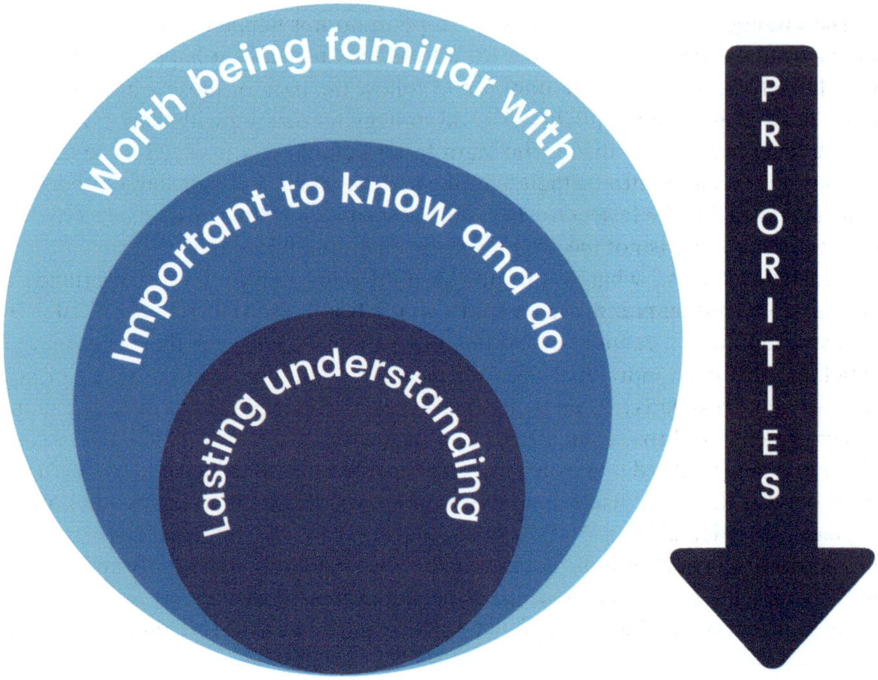

Fig. 4.7 Order of priorities for planning

- What knowledge (facts, concepts and principles) and skills (procedures) will students need to perform effectively and achieve the desired results?
- What activities will equip students with the necessary knowledge and skills?
- What will students need to be taught and trained in, and what is the best way to do this in light of the performance objectives?
- What materials and resources are most useful for achieving these objectives?
- Is the overall design coherent and effective?

These indications, proposed in the book to which I refer, should serve as a guide rather than a mandate. We can incorporate these ideas in a flexible way that is appropriate to our own needs and possibilities but always bearing in mind the central idea of this approach: start by thinking about what we want students to learn in our course.

An alternative approach, but one that is consistent with the aforementioned logic, is the SMART method (specific, measurable, achievable, relevant and time-bound) (Doran 1981). Originally, this approach was developed for project management, however it is very appropriate when it comes to what we are dealing with in this chapter. After all, it is about managing the educational project of a course.

- Specific: concisely state what will be done and who will achieve it. It should be clear and tangible.

- Measurable: provide details of how an action or skill will be measured. A tangible measure, such as a grade or score, can be considered (but is not mandatory if clearly measurable criteria or standards are used).
- Achievable: feasible within the time frame and with the resources available. It should be achievable within the learning environment.
- Relevant: appropriate to the purpose of the class, lesson or program. It should be relevant to the student's learning outcomes.
- Time-bound: with a specific deadline for completion. A realistic time frame should be set for learning to take place.

Chapter 5
About Thinking

> *I cannot teach anybody anything. I can only make them think.*
> *Socrates.*

They say we live in a VUCA world (the acronym for Volatile, Uncertain, Complex and Ambiguous), an idea coined in the nineties, after the end of the Cold War, but one that is still absolutely valid. You may be wondering what this has to do with education, and I would say: EVERYTHING. And I claim that it has EVERYTHING to do with education because we are training people who live and will develop in this VUCA world in which the only certainties are those that come from the basic laws of nature: Newton's laws (motion and gravitation), Maxwell's laws (electromagnetism), the laws of thermodynamics, Einstein's theory of relativity, etc. Almost everything else is, nowadays, subject to review and questioning in line with the new knowledge that is being produced at an astonishing speed in different areas of human activity.

In the book *Critical Path* (Buckminster Fuller 1982), the author describes what he calls the "knowledge doubling curve". He said that, until the end of the nineteenth century, knowledge had doubled every 100 years, but as he was writing the book, this doubling was taking place every 18 months. By the year 2000, it was estimated that this increase was taking place every year; and there are those who say that, nowadays, it happens daily (Sandle 2018). However it is measured, it is undeniable that, nowadays, knowledge is increasing in geometric progression and that trying to include the knowledge of a discipline in the time it takes to complete a university degree would be like trying to put all the water in the sea into a hole, a reference to the well-known anecdote of St. Augustine and the child.[1]

All this introduction leads me to my aim in this chapter, which, as the title suggests, is about thought. As I said before, one of the most important skills we need to develop in order to function in today's society is learning how to learn. Learning how to think is just as important, if not more so. But what is thinking? Can you teach someone to think? Let's take it one step at a time.

[1] It can be consulted at: https://stvincentferrer.com/the-mystery-of-the-trinity/.

U. R. Cukierman, *The Art of Teaching, the Magic of Learning: Towards Innovative Higher Education Systems*, Studies in Systems, Decision and Control 697,
https://doi.org/10.1007/978-3-032-35021-3_5

5.1 Critical Thinking

According to the Dictionary of the Spanish Language, the word think means 'to form or combine ideas or judgments in the mind' (Royal Spanish Academy, n.d.). All living beings think and, through thought, we can reason, solve problems, make decisions, etc. There are different types of thinking, such as critical, creative, analytical, logical, strategic, etc. The different types of thought are not mutually exclusive; we tend to use a combination of them depending on the context and the nature of the problem or task in question. It is not my intention, nor am I qualified, to develop this subject in depth, so I will restrict my analysis to the first type of thought I mentioned: "critical thinking" (CT). Nowadays, it is essential to develop this skill, and teachers can (and I would say should) teach our students to think critically. What is more, before trying to teach it, we should first learn how to do it ourselves. Therefore, in this chapter, I will provide some tools to understand what it is about and learn how to develop, teach and evaluate this fundamental skill for life in modern society.

You should listen to this podcast; it's only 16 min long:

https://bit.ly/Podcast_Marina

Although there is a lot of discussion nowadays about Critical Thinking, it is not really a new concept. Its origins can be traced back to Ancient Greece, to Socrates, Plato and Aristotle. Much closer in time, in 1952, a philosopher and mathematician from Cornell University published a book entitled *Critical Thinking*, thus establishing the name we use today (Black 1952). To understand how certain subjects are related to each other (some would say that everything is related to everything else), it is worth mentioning here that it was also Black who created the concept of "fuzzy logic", which is at the basis of what we know today as "artificial intelligence" (a subject that I will develop in Chap. 7).

CT is directly related to the higher-order thinking skills in Bloom's taxonomy (1956) (see Fig. 5.1).

A practical way of promoting CT is through the so-called "inquiry-based approach cycle" (Bruce and Bishop 2002). It begins by asking questions; then, investigating the possible solutions that will lead to the creation or construction of knowledge, the discussion of experiences and discoveries, and the subsequent reflection on the results; in a recurrent cyclical way, as can be seen in Fig. 5.2.

But let's pause for a moment at the first step of this cycle. In general, traditional education is based on teaching students to answer questions, but, in reality, it is just as important, if not more important, than answering questions correctly to be able to ask relevant questions, and then look for the appropriate answers. Some time ago, at an international conference I attended, a manager from a major multinational company was asked how they selected their staff, to which he replied that one of the most important skills was being able to ask good questions. In more pedagogical

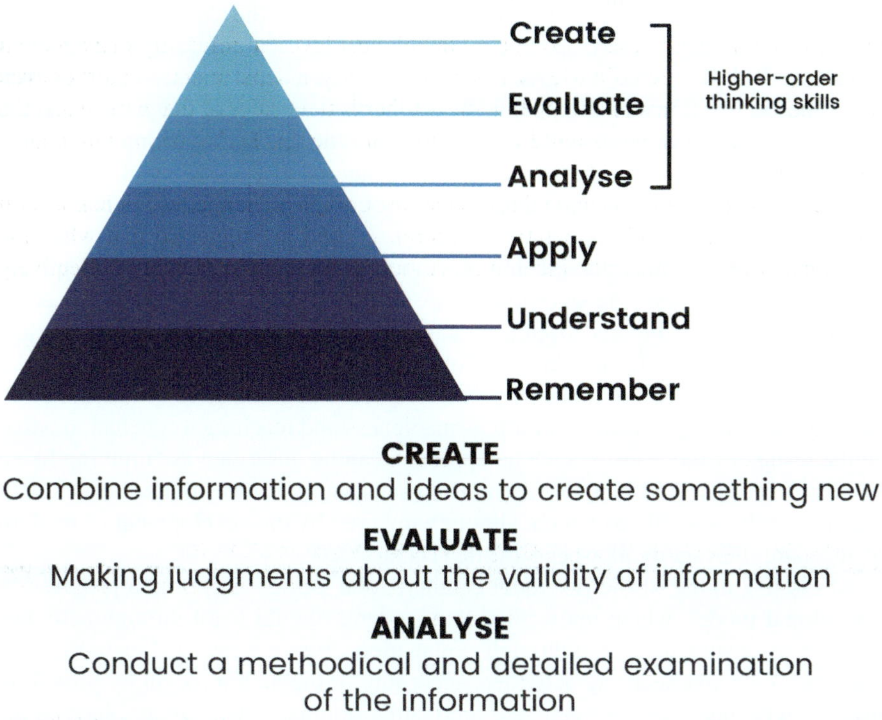

Fig. 5.1 Bloom's Taxonomy

Fig. 5.2 Inquiry-based approach cycle

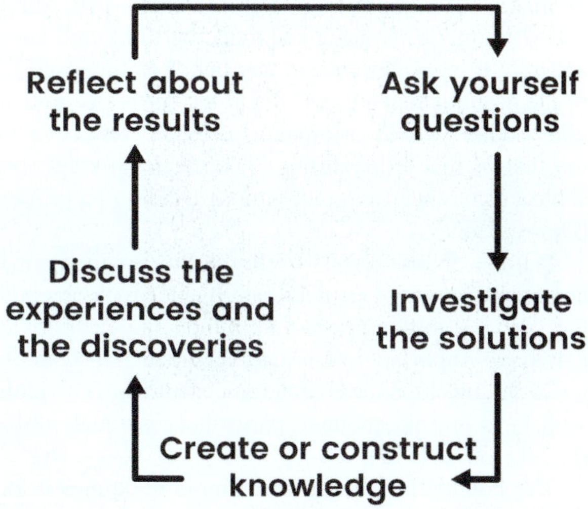

terms, "it is not only teachers who have to ask questions in the classroom. If we want students to continue to develop their curiosity, build critical thinking and reflection and research skills, we need to give rise to their concerns, and also teach them to ask good questions." (Furman 2021, p. 150). Along the same lines, the great Brazilian pedagogue and educator Paulo Freire argues that it is necessary to develop a pedagogy of the question. He tells us that "we are always listening to a pedagogy of the answer. Teachers answer questions that students have not asked" (Freire and Faundez 2013, p. 69).

Critical thinkers must possess and practice the ability to look at a problem from different perspectives. The question is: are we taught or do we feel the need to maintain conditional beliefs, to subject them to honest tests and to be willing to reject them if they do not conform to facts and observations, regardless of the subject under consideration? A critical thinking approach requires not going directly to the answer, but rather proposing it, testing it for reasonableness and reaching a conclusion based on the results of those tests. Such an approach can be described as "thinking like a scientist", but it would be more accurate to say that all critical thinkers, including scientists, rely on methods which, although inspired by the development of modern science, are relevant to all aspects of life (Haber 2020).

At the beginning of the twentieth century, John Dewey proposed a progressive educational model, which maintained that students should learn through activities based on discovery, instead of through explanations from teachers and memorization exercises. Effective learning activities, according to the author, begin by providing students with instances that generate motivating doubts in their minds, such as ill-defined problems without obvious solutions, especially in subjects of interest to individual students. Once that doubt has been raised, it is the teacher's responsibility to channel the students' attempts to dispel that doubt in a logical manner. Dewey called this mode of reasoning "reflective thinking", summarized as "active, persistent, and careful consideration of any belief or supposed form of knowledge in the light of the grounds that support it and the further conclusions to which it tends". While other authors would eventually substitute "reflective" for "critical", Dewey provided the first of many definitions for critical thinking; and all subsequent work on the subject can be seen as a dialogue with ideas first proposed in the book *How We Think* (Dewey 1989).

In times of fake news, like the ones we are living in, in which fictitious realities are created by means of artificial intelligence with objectives that range from promoting the sale of a certain product to influencing elections for government positions, it is extremely important to question statements, instead of accepting their façades. Critical thinking skills have been too absent from curricula, which are often overloaded with facts and procedures, partly because such skills are more difficult to assess (Fadel y otros 2015).

The Foundation for Critical Thinking defines it as "that way of thinking, about any subject, content or problem, in which the thinker improves the quality of their thinking by analysing, evaluating and skilfully reconstructing it" (The Foundation for Critical Thinking).

To learn to think critically, one must evaluate information and arguments critically, trying to see patterns and connections in order to build meaningful knowledge and apply it in the real world (Fullan and Scott 2014).

One of the characteristics of today's world is practically unlimited access to information. Since the advent and massification of the Internet, social networks and, even more so, the emergence of generative artificial intelligence (a topic that I will develop in detail in Chap. 7), it seems that all questions have a single and immediate answer. Riding this wave, we come across what I call *opinators* on a daily basis, serial opinion formers on any topic who base their conclusions on the headlines of digital newspapers, which, deep down, are designed to attract attention and clicks rather than to provide truthful and valuable information. As the well-known Argentine psychoanalyst, writer and thinker proposes:

> …we inhabit the world of opinion. Giving an opinion is easy. Thinking is difficult. Anyone can give an opinion, not everyone thinks. And if someone refuses to give their opinion on a subject about which they do not know enough, they are accused of not wanting to take a chance. As if life were a game in which knowledge does not matter (Rolón 2023, p. 312).

Has it ever occurred to you that the full text of a news item does not match what the title says? Something similar happens with social networks, which reproduce information and news without limit and without the slightest verification of its veracity; and, to make matters worse, selected by their algorithms, in a way that reinforces the cognitive bias that we all have, in one way or another. This is how what is known as the "filter bubble" is created, as described by Internet activist Eli Parser in his book *The Filter Bubble: How the network decides what we read and what we think* (Parser 2017).

This situation means that the development of the ability to think critically is of vital importance today. Let us analyse, then, what an argument is and how it can help us to critically analyse information, and to be able to express informed and reasoned opinions.

5.2 What Is An Argument?

Let's start by saying that an argument is a reasoned attempt to express a particular point of view on a debatable subject. Let's analyse how an argument works through an example.

➡ Activity 4.3: Carefully read the following reasoning by a student:

> A student says that they will not have to study much for a specific subject (claim) because they regularly attend the tutorial classes (reason); and, in the faculty forums, other students say that the teachers are more lenient with those

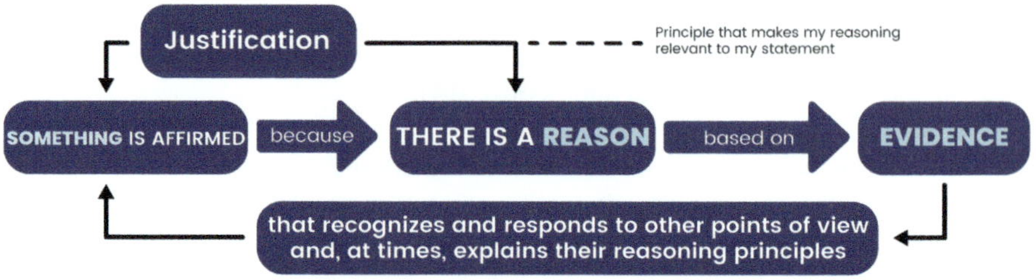

Fig. 5.3 Argument validation process (Booth et al. 2008)

who attend these classes (evidence). Based on this argument, the student makes sure that the teacher notices their attendance at the tutorial classes.

Is this reasoning correct? Before moving on, write on a piece of paper why you think it is or why you think it is not.

Now that you have completed the task, let's analyse a little more in depth what the student said.

What other reasons underlie the claim that the student will not have to study as much if he goes to the tutorial classes? What are the characteristics of the students who attend the tutorial classes? For example, do they understand the subjects better and is that the factor that gives them the advantage? What is it about the tutorial classes that can make such a difference? Could this be achieved in another way? After critically analysing what the student says, we will come to the conclusion that nothing guarantees that, by regularly attending tutorial classes, they will not need to study hard. Therefore, the argument is false. This process can be illustrated as shown in Fig. 5.3.

In order to judge a given piece of evidence, we have to analyse both the quality of the information and the quality of the interpretation of that information. Both to develop our own capacity to think critically and to teach our students how to do so, it is necessary to demonstrate and exercise this procedure in such a way that it becomes a habitual practice. Debates are usually good opportunities to put these strategies into practice. But there is a caveat when it comes to organizing classroom debates: the aim should be to promote dialogue as opposed to discussion. Public debates between candidates for a particular post are often a good example of this, as they generally end up being a discussion rather than a debate. A dialogue is a conversation in which ideas and comments are presented alternatively, but it becomes a discussion when the participants start to defend their own opinions or interests. Let's look in a little more detail at the differences between a dialogue and a discussion (see Fig. 5.4).

A good practice to avoid misunderstandings, promote constructive dialogue and develop CT is what is called the "translation process", whereby the listener presents the logical translation of the argument to the person who presented it, and then asks them if they correctly and honestly understood what they were trying to say before discussing the issue. This process requires empathy, the capacity to put oneself in

Dialogue mean:	Discussing means:
•Be collaborative	•To oppose
•Be understanding	•To want to prove the other person wrong
•Listen to discover the true meaning of what the other person is saying	•To look for flaws
•Reassess your opinion/assumption	•To defend one's own opinions/
•Be open-minded	•assumptions/beliefs
•Set aside your own beliefs and prejudices	•To be close-minded
•Look for validity in all ideas	•To look for weaknesses in the ideas of others
•Respect all participants	•To treat others badly
•Explore different possibilities	•To believe you have the right answer
•Dialogue has no set ending	•Discussion demands a conclusion

Fig. 5.4 Dialogue versus discussion

the other person's place to discover what they believe and why they believe it. In addition to facilitating more honest dialogues, that empathy also happens to be a powerful check on "confirmation bias", which is the tendency of the human mind to accept what fits in with what we already believe and reject what does not, a tendency exploited by those who design social media algorithms. It is a flaw in our reasoning that makes us vulnerable to misunderstanding and manipulation.

You should watch a video on this subject on YouTube:

https://bit.ly/Pens-Critico

5.3 How Can We Teach Critical Thinking?

So far, I have developed concepts that describe what critical thinking is and what the elements and strategies involved in it are. Although it is possible (and perhaps necessary) to include a subject on critical thinking in the curriculum, I believe it is much more convenient to develop this competence in a cross-curricular way in different courses and integrated into the learning of the concepts specific to each subject. This is how Ennis understands it, who, as opposed to what he calls the "general approach" (that is, teaching critical thinking skills and dispositions separately from presenting the content of existing subjects), proposes "infusion" or "immersion" (Ennis 1989). On the one hand, "infusion" is a deep, reflective and well-understood instruction in which students are encouraged to think critically about the subject they are studying, and in which the general principles of critical thinking dispositions and skills are made explicit. On the other hand, "immersion" is a similar type of thought-provoking instruction in which students are immersed deeply in the subject, but in which the general principles of critical thinking are not made explicit. In any case, it all starts with encouraging students to ask questions; and, based on those questions, they go through the cycle of the inquiry-based approach, described in Fig. 5.2. In practice, this could be done in the following way:

1. Start the class by presenting students with an open-ended question

These types of questions are the ones that may have more than one correct answer. It is important that students understand that there are no wrong answers; if anything, there are opportunities to learn. In fact, incorrect answers (if they were the result of a type of creative, unstructured thinking) should be encouraged at this stage, making it clear that, at some point in the process, the correct answers will come naturally and logically. It is important to allocate a reasonable amount of time for this part of the process, thus allowing students to think about the topic in question; and, in addition, to encourage the greatest possible participation, preventing a few students from monopolizing the interventions.

2. Analyse the different responses received

List the different answers proposed by the students and analyse the reasons for and against each of these answers. Here too, we must avoid introducing our own bias in one direction or another, as we know the correct answer(s). Similarly, we must try to avoid giving extra credit to those who give correct justifications or penalizing those who do not. Finally, in this part of the process, try to get students to make connections with other subjects or with real-life situations so that they move towards the correct answers themselves, placing special emphasis on verifying the sources of information used.

3. Construction of knowledge

Based on the results of the previous step, with the appropriate guidance of the teacher who knows what they want to be obtained from the process in terms of the expected learning outcomes, and following the process described in Fig. 5.3, valid and necessary conclusions should be reached.

4. Finally, the experiences and discoveries will be discussed, the results will be reflected upon in order to restart the process again with a new question, and so on

Richard Felder in his book (Felder and Brent 2016), citing (Fogler y otros 2014), proposes a structured critical reasoning for the critique of a document or presentation, which consists of students doing the following:

- Summarize all the statements made by the author (or the presenter).

- Examine and evaluate the author's reasons (justifications) and gather external evidence (facts, observations, research data, scientific and logical principles) that support or reject the conclusions.
- Identify and evaluate the author's stated and unstated assumptions and biases.
- Draw conclusions about the validity of the author's statements.

In addition to the methods mentioned so far, many authors have developed different techniques and procedures for teaching critical thinking. In this sense, I recommend accessing what its authors call the wheel of reason, published on the website of the Foundation for Critical Thinking (The Foundation for Critical Thinking). This interactive resource provides elements for analysing thought through the identification and questioning of its elementary structures.

Another very valuable resource is offered by the Kialo application, which is available in a free version for educators (Kialo Edu, available in English and other languages). This resource allows users to start a discussion by posting a point of view on a topic of their choice. Other users then contribute to the thread by posting well-structured pros and cons (or "statements") in response to the original statement. Those who access it later can view the discussion history to learn more about a topic and then give their opinion.

Last but not least, I would like to recommend the website developed by the Harvard Graduate School of Education, called "Project Zero", which offers a huge variety of resources for CT development.[2]

5.4 How to Assess Critical Thinking?

The assessment of critical thinking is a widely studied topic in the workplace. In fact, there are tools developed by specialized companies to assess this competence in the process of personnel selection and promotion. One of the best known is the one named after its authors (Watson & Glaser) and which is currently offered by the company Pearson.[3]

In the academic environment, teachers can use their own instruments; and, as I explained in the chapter on Assessment, rubrics are also excellent for this purpose. The Organization for Economic Cooperation and Development (OECD) has developed a rubric that can be very valuable as input for the elaboration of our own rubrics.[4]

In an interesting piece of work (Reynders y otros 2020), the rubric I include as Fig. 5.5 is presented.

---

[2] It can be consulted at: https://pz.harvard.edu/thinking-routines.

[3] It can be consulted at: https://www.talentlens.com/recruitment/assessments/watson-glaser-critical-thinking-appraisal.html.

[4] It can be consulted at: https://web-archive.oecd.org/2021-10-25/614045-class-friendly-assessment-rubric-critical-thinking-es.pdf.

Critical Thinking	Evaluating, analyzing, and/or synthesizing relevant information to form an argument or reach a conclusion supported with evidence.

Category	0	1	2	3	4	5
Evaluating		**Minimally** determined the relevance and reliability of information that might be used to support a conclusion or argument		**Partially** determined the relevance and reliability of information that might be used to support a conclusion or argument		**Extensively** determined the relevance and reliability of information that might be used to support a conclusion or argument
Analyzing		**Inaccurately** interpreted information to determine meaning and to extract relevant evidence		Interpreted information to determine meaning and to extract relevant evidence **with some errors**		**Accurately** interpreted information to determine meaning and to extract relevant evidence
Synthesizing		**Inaccurately** connected or integrated information to support an argument or reach a conclusion		Connected or integrated information to support an argument or reach a conclusion **with some errors**		**Accurately** connected or integrated information to support an argument or reach a conclusion
Forming Arguments (Structure)		Made a claim and provided **incomplete** evidence to support it.		Made a claim and provided **partial** evidence to support it.		Made a claim and provided **complete** evidence to support it.
Forming Arguments (Validity)		The claim, evidence, and reasoning were **minimally** consistent with accepted disciplinary ideas and practices		The claim, evidence, and reasoning were **partially** consistent with accepted disciplinary ideas and practices		The claim, evidence, and reasoning were **fully** consistent with accepted disciplinary ideas and practices
Comments:						

Fig. 5.5 Rubric by Reynders et al. for the assessment of CT

Finally, (Felder and Brent 2016) propose an example rubric for assessing students' critiques of an article or essay (see Fig. 5.6). It is important to note that, in the last element of this rubric, students are asked to reflect on what they did well and what they could do to improve next time. This type of reflective self-assessment (a form of metacognition) is an important step in the development of strong critical thinking skills.

I think it is important to recommend that teachers develop their own rubrics, adapted to the type of activities they organize and the scenarios in which they take place; that they test them and improve them based on their own experience. A tool that may be useful for one teacher in one course or subject may be completely inappropriate for another colleague.

What I have described so far on this subject, similarly to what happens in other chapters, is not intended to exhaust it or cover it in all its richness and potential. Helping our students develop CT is, in my opinion, the best tool we can offer them to face the world we live in today. Moreover, in view of the advent of generative artificial intelligence, which I will refer to in Chap. 7.

Criterion	Max. Points	Score	Comments
Understanding of document content (25%)			
Summarized author's purpose, main ideas, and conclusions	15		
Fairly represented author's viewpoint	10		
Critical analysis (40%)			
Identified author's assumptions and biases	10		
Identified strong, supported, and persuasive arguments and conclusions	10		
Identified weak and unsupported arguments	10		
Stated and persuasively justified a conclusion about the validity of the author's position	10		
Presentation (25%)			
Neat and well formatted	5		
Used correct grammar and good style	10		
Wrote clearly and persuasively	10		
Reflection (10%)			
Identified strong and weak points in this critique (including self-bias) and steps that could strengthen it	10		
Total Score	**100**		

Fig. 5.6 Felder and Brent rubric for assessing student critiques of an article or essay

Chapter 6
About Technology

> *Any sufficiently advanced technology is indistinguishable from magic.*
> *Arthur C. Clarke.*

I want to start this chapter by reproducing an anonymous story that is circulating on the Internet and that I usually use in my workshops on educational technology for teachers. It takes place in any classroom in a school that a supervisor is visiting.

> On his tour of the school, the supervisor observed a chaotic scene in one of the classrooms. The pupils were shouting and the teacher was "entrenched" behind her desk. He then decided to enter the classroom in question.
> "Excuse me. I'm the supervisor. Is there a problem, teacher?"
> "Thank the Lord you came! I'm overwhelmed, I don't know what to do with these kids. The Ministry doesn't provide me with teaching materials. I have nothing to show them or to tell them."
> The supervisor, a teacher at heart, saw a cork on the floor, picked it up and addressed the students.
> "What is this?"
> "A cork, sir."
> "Good. Where does the cork come from?".
> "From the bottle, sir."
> "From the cork oak, from a tree," said another student.
> "From wood," added a third.
> "And what can you make with wood?" asked the supervisor.
> "Chairs, tables, boats," other voices replied.
> "Good!" said the supervisor. "A boat! Who wants to draw a boat? Who will make a map on the board and mark the nearest port for our little boat? Write down which region it belongs to. Which is the other nearest port? What does this region produce? Which poet do you know who was born there? Does anyone remember a song from this place?"

U. R. Cukierman, *The Art of Teaching, the Magic of Learning: Towards Innovative Higher Education Systems*, Studies in Systems, Decision and Control 697,
https://doi.org/10.1007/978-3-032-35021-3_6

And so began a task of geography, history, music, economics and literature.
The teacher was impressed.
"Sir, I will always remember what you taught me today. Thank you so much!!!"
Time passed. The supervisor returned to the school and looked for the teacher.
She was in the same situation as before.
"Miss, what's wrong? Do you remember me?"
"Yes, sir. How could I forget? I'm so glad you're back! I can't find the cork. Where did you leave it?"

We can compare the cork in this story to technology. What is important is not the technology, but the use we make of it in terms of learning objectives. A hammer can be used to build something or to destroy it. Technology is a tool that must be used correctly. This is the topic I will be discussing in this chapter.

6.1 Historical Introduction to Educational Technology

The history of educational technology shows that an invention over 200 years old is still, perhaps, the most influential technological innovation of our time. Indeed, Josiah F. Bumstead pointed out in an essay in 1841 that the blackboard is a revolutionary technological invention (Krause 2000). Numerous scholars have agreed with the author, who also stated:

> The inventor or introducer of the black-board system deserves to be ranked among the best contributors to learning and science, if not among the greatest benefactors of mankind; and so he will be regarded by all who know its merits, and are familiar with school-room trials. (Bumstead 1841, p. 71)

Much closer in time, in one of his memorable lectures which I had the great opportunity to attend, Litwin (2002) pointed out, with regard to the incorporation of technologies in the classroom, that the first tool that helped teachers solve the problems of difficult comprehension and teaching was the modest chalk and blackboard: "That is a technology that, I still think, is cutting-edge" (p. 6). On this point, a quote from Buckingham is pertinent (2008), when he announces that:

> It is as inevitable today to use different types of technology in education as it always was: the book, the pencil and the blackboard are technologies, as are the computer, the video recorder or the latest mobile communication device. The question is not whether to use technology, or even which technologies to use, but why and how to use them. The polarization that characterizes this debate — the endless confrontation between "supporters" and "opponents" — has made it very difficult to address the fundamental issues. (p. 224)

As this author points out, the process of incorporating technologies into education did not begin with the blackboard nor did it end there. From oral communication (passing through the invention of writing and, later, that of the printing press) to the more recent, in relative terms, use of computers and the Internet, the idea of using technological tools to support teaching and learning is a constant that has endured (and will continue to endure), of course adapting to the times, uses and customs. In this historical evolution, we can consider that the emergence of digital technologies constitutes a milestone that changed the way we live and also the way we educate. Today it is hard to imagine everyday life without a computer (or a cell phone) and an Internet connection. In fact, the inventor of the hypertext markup language (HTML), better known as "the inventor of the Web," has recently stated that "access to the Internet should be a universal right." (Berners-Lee 2020). It is precisely these digital technologies that, a couple of decades ago, burst into the classroom uninvited, in fact, with no little resistance from teachers and educators. An example of what I am proposing here is the argument of the researcher Larry Cuban, who said that "computers are, to a large extent, incompatible with the requirements of teaching and that, for the most part, teachers will continue to reject their use as tools for students to work with during class" (Becker 2000, p. 3). At a local level, there were also those who tried to oppose this trend: in 2006, the Director General of Culture and Education of the Province of Buenos Aires banned the use of mobile phones by teaching staff and students who were carriers or users within the school environment and during class time throughout the entire education system (Dirección General de Cultura y Educación de la Provincia de Buenos Aires 2006). This regulation was repealed 10 years later. In the meantime, countless investigations were carried out which proved that, when there are clear objectives and adequate preparation, the use of cell phones for educational purposes is extremely convenient (Cukierman 2005; Cukierman et al. 2007, 2017, 2018).

6.2 Educational Technology in the Field of Higher Education

In my doctoral thesis, completed and approved at the end of 2022, I identified a series of emerging technologies based on different sources that, according to their criteria, are (and will be) of great impact on higher education:

- artificial intelligence
- extended reality
- security and privacy
- digital transformation
- personalized learning

Due to the great impact it has had in recent times, I will develop the first of these in a special chapter later on. For now, we will move on to the other four.

6.2.1 Extended Reality

The concepts of augmented reality and virtual reality have been used for some time in certain applications such as commerce, medicine, tourism and entertainment, but they have only recently started to be used in educational applications (Antonioli y otros 2014). Currently, people talk about mixed reality (see Fig. 6.1), thus including technologies that combine augmented reality tools with virtual reality; or, as researchers Milgram and Kishino (1994) proposed 30 years ago, "environments that do not necessarily involve total immersion and complete synthesis, but that fall somewhere on a continuum of virtuality" (p. 2).

Thus, we arrive at the concept of extended reality, "a general term that covers a wide range of technologies along a continuum, with the real world at one end and fully immersive simulations at the other" (Pomerantz 2018, p. 6).

In recent years, due to the popularization and greater ease of access to the necessary technological tools, the application of extended reality in educational settings has been growing (Cabero Almenara and Puentes Puente 2020; Mota et al. 2016; Peng Chen 2016). Extended reality involves an adaptation to the new hypertextual culture, specific to the new generations; and it makes it easier for students to navigate, interact and construct their own knowledge based on the use of different resources and symbolic systems. It allows the visualization of an object or phenomenon from different perspectives, in a way that enhances the spatial intelligence of students.

Therefore, the possibilities of manipulating the digital object allow us to trace 3D scenarios from reality itself, capturing space and movement in a way that is similar to the human experience. Thus, when coming in contact with experiences that simulate real scenarios, students can recognize information, formulate hypotheses and risk answers, turning extended reality into an excellent resource for promoting short- and long-term memory.

Finally, it is worth mentioning the idea of the metaverse (or metauniverse), which has gained considerable popularity recently. The term metaverse, coined in the science fiction novel *Snow Crash* (Stephenson 1992), describes a persistent and immersive 3D virtual environment in which a user anywhere in the world with access to a computer could carry out all kinds of activities, from business to entertainment. This concept refers to the combination of certain technologies already mentioned in this section, such as extended realities, complemented by so-called "virtual worlds"

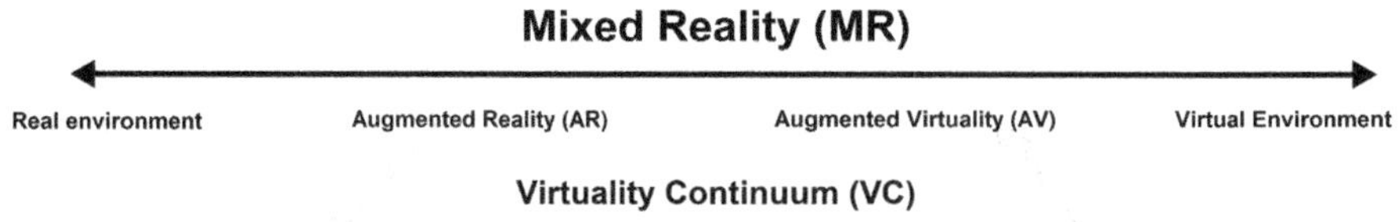

Fig. 6.1 Simplified representation of a virtual continuum

(such as Second Life[1] or Decentraland[2]), which thus becomes a virtual space in which the real and the digital converge.

With regard to its application in education, although there are several experiences (Anacona et al. 2019; Duan et al. 2021), this concept still has little practical application, especially in Argentina.

6.2.2 Security and Privacy

On the face of it, these issues appear to be unrelated to education, but what is in fact well known is that, as soon as you start surfing the Internet, your personal data and information become subject to the resources used by companies and public and private organizations to collect, store and process them for a wide range of purposes; not always ethically acceptable and sometimes illegal. This situation is exacerbated by the use of artificial intelligence. Therefore, not considering or minimizing the influence of this aspect when studying the applications of information and communication technologies (ICT) in education would be a very risky attitude, equivalent to driving a car without considering the risks involved in driving on streets and highways where other vehicles and people are moving simultaneously.

Now, what aspects of security and privacy should we consider in this analysis? To answer, we need to define four basic concepts (Stallings 2014):

- Privacy: only authorized entities are required to have access to the information; this type of access includes printing, viewing and other forms of disclosure, in addition to simply making the existence of an object known.
- Integrity: the data is required to be modified only by authorized parties.
- Availability: the data is required to be available to authorized parties.
- Authenticity: a computer or service is required to be able to verify the identity of a user.

Therefore, these issues should be considered as a priority in the analysis and decision-making regarding the adoption of technologies in higher education. In a recently published report (EDUCAUSE 2021), it is stated that data security and privacy are of extraordinary and growing importance on the horizon of higher education institutions; this situation has increased significantly during and since the pandemic due to the increased use of resources such as videoconferencing and various Internet platforms. In the aforementioned report, it is stated:

> With the dramatic shift to remote work and remote learning in 2020, threats are at an all-time high. The introduction of contact tracing, the increased use of home personal devices, and problems related to the proliferation of videoconferencing have ushered in unprecedented concerns. (EDUCAUSE 2021, p. 4)

[1] It can be consulted at: https://secondlife.com.

[2] It can be consulted at: https://decentraland.org.

6.2.3 Digital Transformation

Digital technologies have been penetrating all human activities to an increasing degree since the appearance of the first digital computers, approximately 80 years ago. Nowadays, they constitute the lifeblood that flows through almost all the devices and networks used in everyday life. This fact could be considered a digital transformation, but this concept refers to something much more specific. It is "a process that aims to improve an entity by activating significant changes in its properties through combinations of information, computing, communication and connectivity technologies" (Vial 2019, p. 3). Another way of defining it, from a point of view closer to higher education institutions, is as "a series of profound and coordinated changes in culture, workforce and technology that enable new educational and operational models and transform an institution's operations, strategic directions and value proposition" (Grajek and Reinitz 2019).

Once again, the question arises as to the relationship of this concept with higher education. Although, based on everything expressed so far, that relationship may seem obvious, I will try to substantiate this assertion in a reasoned and justified manner below.

Digital transformation in universities has two fundamental dimensions (Llorens and Fernández 2018): the ability to change or create new processes, and the capacity to create processes of a strategic nature.

Based on these two dimensions, the grid of digital transformation is designed, establishing different situations in which a university can find itself in relation to the impact of ICT:

- Digital management: technologies help to make existing processes more efficient, even if they are not strategic for the university.
- Digital innovation: technologies enable the creation of new processes or transform existing ones in a disruptive way, but without turning them into strategic processes for the university.
- Digital government: technologies are applied to existing processes, but which, when properly governed, become strategic for the university.
- Digital transformation: the high potential of new technology leads to the creation of new disruptive and strategic processes for the university.

Figure 6.2 shows the grid of the digital transformation.

From this analysis, the importance of digital transformation for the management of higher education institutions is clear. This factor could be assumed to be unrelated to pedagogical and didactic issues, but on the contrary, it significantly influences them, facilitating or limiting them. In effect, poor management of digital tools will limit what teachers and students can do to improve the teaching and learning processes.

In short, a higher education institution undergoes a true digital transformation when the curriculum and the organization of teaching and learning change to take advantage of available technologies.

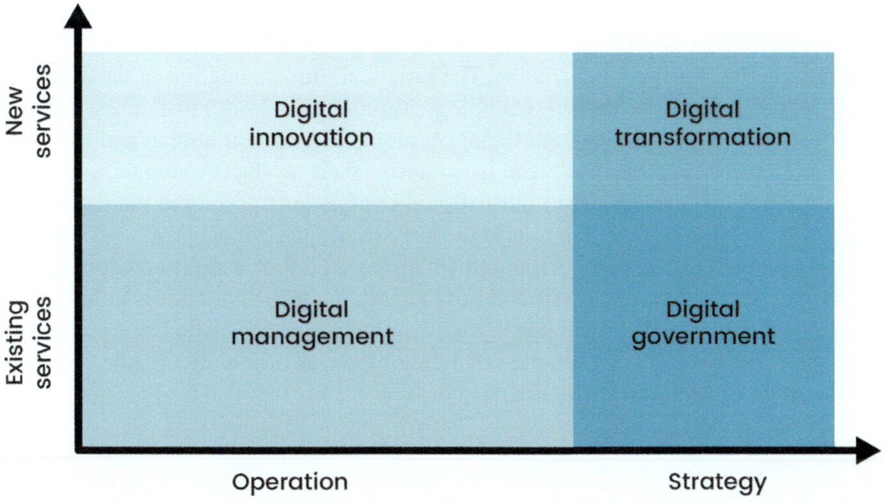

Fig. 6.2 Digital transformation in universities. *Note* Taken and translated from Llorens and Fernández (2018)

6.2.4 Personalization of Learning

Throughout human history, there have been different ways of organizing the education of people, and in particular of young people. The oldest were aimed exclusively at nobles and the clergy, and took place within the family, or in temples and churches. Later, between the mid-eighteenth and early nineteenth centuries, the need to educate all children arose. This change emerged in Europe, fundamentally due to the demands of the Industrial Revolution and the political and social reorganization of the era (Cukierman et al. 2018). From that time onwards, mass and compulsory education for all young people has been promoted and encouraged. This massification led, in one way or another, to the standardization of content and strategies. It was not until the twentieth century that educational strategies that consider and respect the individuality of people began to be promoted. In this sense, the contributions of Pierre Fauré in France and Víctor García Hoz in Spain are noteworthy. (Calderero y otros 2020). More recently, Gardner's contributions regarding multiple intelligences have also had a great influence on this conception (Gardner 1983, 1991; Gardner and Hatch 1989).

This trend towards the personalization of learning, originating in pedagogical and sociological approaches, coincided with the dissemination and widespread use of digital technologies in education, leading to the development of platforms and tools that, from their conception, aim to facilitate this personalization. Such is the case of personal learning environments (PLE), which emerged at the beginning of the twenty-first century as a consequence of Web 2.0, and which have undergone great development, especially in higher education and continuous learning. (Torres Kompen y otros 2019). Finally, along the same lines, these systems have been incorporating the aforementioned artificial intelligence techniques with the aim of systematizing and automating this personalization. The practical implementation of these innovations requires highly advanced technological tools, which, in most cases, can only be developed by companies with the necessary and sufficient human and economic resources; therefore, they will need to monetize their products to obtain a return on investment. This has led large multinational companies to promote the use of their platforms. All this analysis leads us to the need to adopt a critical or, at least, cautious position on the issue at hand. Although we understand the need for and convenience of personalizing learning, we should not fall into the trap of adopting technological solutions that may not be available to many higher education institutions and, even if they were, this would not guarantee that the objectives originally intended with this personalization would be achieved.

6.3 A Few General Points on the Subject

As I described at the beginning of this chapter, there has always been some kind of technology involved in educational processes, whether traditional or innovative, analogue or digital. They were and are tools that, like a chisel in the hands of a sculptor, can produce wonderful or forgettable results. The difference between the two results lies in the hands of the person using the tool, in their skill, their training, the passion and interest they put into the task, their commitment to their art: in our case, as teachers, to the art of teaching.

When digital technologies appeared, and with them the rise of computing and the Internet, some people called them "new" technologies, but they very quickly lost their novelty due to the emergence of more modern ones. And this continues to happen without interruption. As Alan Kay, one of the pioneers of devices that are now so commonplace such as notebooks and tablets, used to say, "technology is everything that was not around when you were born". For this reason, today I think it is more convenient to talk about "relevant technologies", that is, those that are selected and used by a trained teacher, in a given environment and for a specific purpose related to the expected learning outcomes. Thus, from modest chalk to AI, they can serve the educational objectives of a suitable and dedicated teacher.

A decade ago, Brynjolfsson and McAfee postulated:

> We can change the way we deliver education by putting to work digital technologies that have been developed over the past decade or two. The good news is that compared to other industries such as media, retailing, finance, or manufacturing, education is a tremendous laggard in the use of technology. That's good news because it means we can expect big gains simply by catching up to other industries. Innovators can make a huge difference in this area in the coming decade. (Brynjolfsson and McAfee 2014, p. 210)

I do not think these authors could have imagined the technological advances that have taken place since then, but what they wrote in their book is as relevant now as it was then, which is both good and bad news. The bad news is that education is still very far behind in the use of technology. The good news is that we can still expect great benefits if, once and for all, actions are taken at the governmental, institutional and personal level to make this happen; and, at the center of these actions, should be teacher training.

> The integration of technology into teaching practice is no longer an option to choose from, it is installed, against all resistance in the subjectivity of the people who inhabit the present times both inside and outside institutions. Openness and vanishing points are constructed in the way in which this inclusion is approached, with its critical possibilities and openness to dialogue. (Lion y otros 2023)

There is no doubt, then, that the debate over "technology yes or no" is an anachronistic one and that, as I usually say in my workshops and courses, education is not about knowing what the technology of the future will be, but rather how technology can help us create the education of the future… Or should I say, "the education of the present"?

To reinforce everything expressed so far, I quote the person who currently directs the UNESCO International Institute for Higher Education in Latin America and the Caribbean (IESALC):

> … there is an undeniable impact of technology on higher education, but […] the predominant uses tend to reinforce traditional strategies. The problem does not lie in the availability of resources or applications, but in the lack of mechanisms and incentives for pedagogical redesigns, some of which could be enhanced by the intensive use of technology, to be considered a priority by teachers. It is more of a systemic than a technical problem. (Pedró 2023)

Finally, when talking about technology, it is important to consider the students' perspective. Our students do not think in terms of technology, but in terms of the activities enabled by that technology. When asked about technology, they only think about the technologies they currently use. For them, the activity they carry out is more important than the technology that allows them to carry out that activity; in other words, the how is more important than the what. Therefore, as teachers we must take this fact into account when choosing the technologies we will use. More technology does not necessarily mean better education, as many national plans for the mass distribution of technology intended, which may have been very positive in terms of improving access, but which did not impact on the quality of learning. For this reason, technology should be focused on facilitating the learning process, and its use should be carefully planned. Using technology without objectives, without plans or without adequate teacher training can lead to poor results.

In response to the question that is repeated every time a new technology appears in the classroom, which is whether teachers can be replaced by it, my answer is that there are no substitutes for good teachers, since it is not just a question of choosing which technology to use, but knowing why and how to apply it. That is our task as teachers.

Chapter 7
About Artificial Intelligence

Artificial intelligence is neither artificial nor intelligent.
Kate Crawford.

Below I reproduce, with a few modifications, a text I wrote recently entitled "Artificial Intelligence vs. Natural Intelligence: A dilemma for education?"

Toward the end of 2022, the company OpenAI made the ChatGPT application available for public use. ChatGPT is a chatbot based on artificial intelligence. Since then, a kind of fever has been generated that seems to affect all fields of human activity, especially work and education. However, this is not a new issue at all, and its implications have been analysed and evaluated for a long time. In fact, in 1950, the brilliant Alan Turing, one of the fathers of computer science and a forerunner of modern computing, was already asking whether machines could think. In this chapter, I will briefly describe the evolution of this topic over the years and the associated concepts; and, finally, I will make a critical analysis of its potential impact on education.

7.1 History and Concepts Involved

As I mentioned in the introduction, Alan Turing was the first to raise the idea that a machine might be capable of thought; in fact, in the last paragraph of that article, he puts forward the following hypothesis: "We may hope that machines will eventually compete with men in all purely intellectual fields" (Turing 1950). A few years later, an article was published in which the concept of artificial intelligence (AI) was mentioned for the first time. This work is considered the cornerstone of this new field. In it, the authors refer to AI in the following way:

> An attempt will be made to find how to make machines use language form abstractions and concepts solve kinds of problems now reserved for humans and improve themselves. [...] For the present purpose the artificial intelligence problem is taken to be that of making a machine behave in ways that would be called intelligent if a human were so behaving (McCarthy y otros 1955).

U. R. Cukierman, *The Art of Teaching, the Magic of Learning: Towards Innovative Higher Education Systems*, Studies in Systems, Decision and Control 697,
https://doi.org/10.1007/978-3-032-35021-3_7

Since then and up to the present day, a wide variety of technologies and applications that use or are related to AI have been developed. It is not my objective here to give a detailed description of them, but it is worth mentioning some concepts that have emerged over the years. One of them is machine learning (ML), which was first described by Arthur Samuel in 1959 (Samuel, 1959); and is understood as the use and development of computer systems that have the capacity to learn and adapt without following explicit instructions through the use of algorithms and statistical models to analyse and draw inferences from patterns in the data (Oxford English Dictionary). Another concept is that of deep learning (DL), which is a subset of ML methods based on artificial neural networks. "These methods have dramatically improved the state-of-the-art in speech recognition, visual object recognition, object detection and many other domains such as drug discovery and genomics" (LeCun et al. 2015). To conclude this brief conceptual review, we should mention big data (BD), which refers to the handling of large volumes of information facilitated by the accessibility and connectivity provided by the Internet.

> The Internet of Things (IoT) is a by-product of this connectivity and a key factor in the proliferation of data. The connectivity between devices and between humans and devices generates a considerable amount of data, never before imagined, which represents an opportunity in terms of its analysis and associated decision-making (Cukierman and Vendrell Vidal 2020).

This story includes the concepts of generative AI (GAI), which consists of the application of AI to create content such as text, images, sounds, video and even computer code; and large language models (LLM), which are ML models that can understand, predict and generate human language. Finally, as a result of the application of these techniques, the concept of a generative pre-trained transformer (GPT) emerges. ChatGPT, mentioned in the introduction, is an LLM-based chatbot developed by the company OpenAI[1] and publicly presented on November 30, 2022. It was designed to understand and generate text in natural language conversations, covering a wide range of topics and applications.

Figure 7.1 shows a graphic representation of the relationship between all the concepts described so far.

In short, we can say that AI is a field of computer science that tries to simulate how human beings think. The system is fed data to process and create trained models that use that data as a reference; and the more data it has, the better it can learn. Once the process is complete, the AI can be asked questions and asked to make estimates and act based on what it has learned. The scope and accuracy of the AI response depends mainly on the quality of the training data and the algorithms. This operating logic can be represented as shown in Fig. 7.2.

Considering the background I have described above, one might wonder why the topic has gained so much popularity now, when it is nothing new. The ChatGPT system itself was consulted, and this was the answer it offered:

---

[1] It can be consulted at: https://openai.com/.

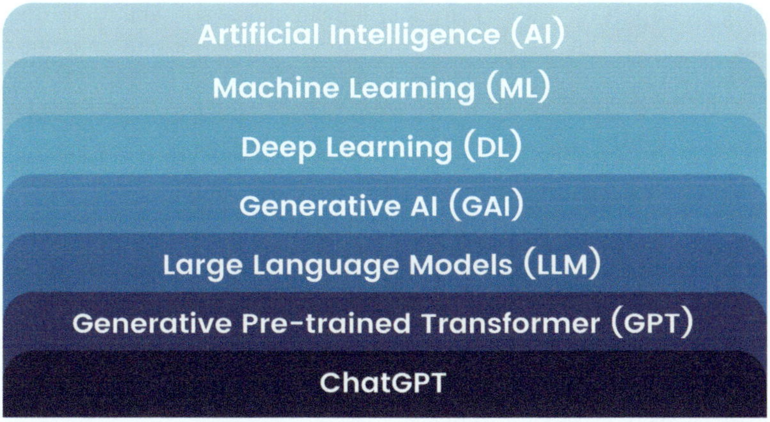

Fig. 7.1 Relationship between concepts (Adapted from Querci della Rovere 2023)

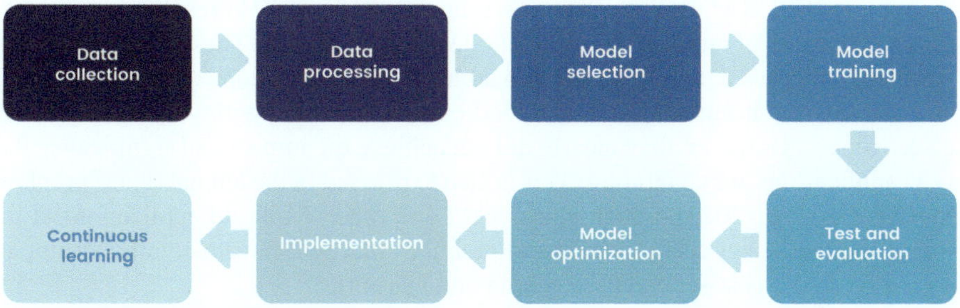

Fig. 7.2 How AI works

> Although artificial intelligence tools have been on the market for a long time, the impact of models such as ChatGPT is usually attributed to the combination of their advanced capacities, their scale and the way in which they have been made accessible to a wide audience, encouraging creativity and the exploration of possible applications.

Another answer to this question can be found in a more traditional source, such as an article published a few days after the launch of ChatGPT in the prestigious Harvard Business Review:

> While versions of GPT have been around for a while, this model has crossed a threshold: It's genuinely useful for a wide range of tasks, from creating software to generating business ideas to writing a wedding toast. While previous generations of the system could technically do these things, the quality of the outputs was much lower than that produced by an average human. The new model is much better, often startlingly so. (Mollick 2022)

Without a doubt, this is a technology that has revolutionized the world. Proof of this is that ChatGPT reached 100 million users in just two months, which is very striking compared to the nine months it took for the TikTok application or the two and a half years for Instagram (Hu 2023). Therefore, it is worth taking it into account and analysing its impact on different areas of human endeavour; and, in this particular case, on education. The renowned Israeli historian and writer said about the subject: "AI has hacked the operating system of human civilization" (Harari 2023b).

7.2 Artificial Intelligence and Education

The use of AI in education is nothing new either. Concepts such as learning analytics and adaptive learning have already been in use for 20 years.

> The first of these derives directly from what is called "Business Intelligence" and also from what is known as "Web Analytics". In both cases, it is about collecting and analysing large volumes of information to facilitate decision-making and/or understanding user behaviour (Cukierman and Vendrell Vidal 2020).

Much research has been conducted in recent years on the effectiveness of the aforementioned techniques. A search of the database of the Institute of Educational Sciences in the USA[2] demonstrates this, returning 815 and 1941 results, respectively, for the concepts of adaptive learning and learning analytics. These techniques have generally been used, incorporated into educational applications or virtual teaching and learning environments such as Moodle or Blackboard, with varying degrees of effectiveness. However, they clearly did not achieve the impact and significance that GAI applications are generating. This is also expressed in the annual report prepared by EDUCAUSE (2023), which places GAI and other AI-based applications at the top of the list of key technologies and practices that will have a significant impact on the future of post-secondary teaching and learning.

However, when analysing the initial reactions of the educational world, in particular of teachers and institutions, to the emergence of ChatGPT and similar tools, what we see is fear, even panic, as clearly described in dozens of recent publications (Heaven 2023; Mollick and Mollick 2023; Bjork 2023; García-Peñalvo 2023). This reaction is generated as a consequence of the rapid adoption of this tool by students, who find in it a way to do their schoolwork effortlessly and obtain results that meet the standards required for approval. But this attitude is nothing new. Similar behaviour was evident with Internet access, with the use of Wikipedia or with various websites such as rincondelvago.com or monografías.com. Of course, now the ease of use and the results far exceed those of the previous resources. As a matter of fact, the emergence of new technologies has always generated this type of reaction in educational settings. I clearly remember the stir caused by the appearance of electronic calculators in the seventies or cell phones in the nineties: the initial reactions were always rejection and prohibition. It seems that we have not yet fully understood that the problem is not the tool, but its relationship with the content and with the project that gives it meaning (Educar 2013). In general terms, teachers' concerns relate to the possibility of students "cheating" when being evaluated through the use of technology, whether with a calculator, with a telephone or, as was done when digital

---

[2] https://eric.ed.gov/.

technologies did not exist, using the most sophisticated cheat sheets, which could be described as "analogue technology". In short, "the problem is not the technology, but the use that is made of it" (Palazzi 2015). And assessments seem to be the "Achilles heel" of education, since it is nothing more than an instance of verification, instead of an opportunity to continue learning, using technology in the most appropriate way.

> If we want to continue educating when we assess and for students to continue learning, then assessment has to stop being a verification process to see if what we teach appears to have been learned and become a proposal that generates considerations for improving our practices (Maggio 2018, p. 93).

At this point, it is critical to analyse the ultimate meaning of the teaching task, which is to prepare students "not for a life of trials, but for the trials of life" (Costa and Kallick 2004, pp. 173, cited by Furman 2021). Returning to the subject of AI, we should ask ourselves what contributions this technology can make to the development of thinking skills, which are characteristic of human beings and, particularly, of critical thinking. According to Jenna Lyle, an official of the New York City Department of Education, "while the tool may be able to provide quick and easy answers to questions, it does not build critical-thinking and problem-solving skills, which are essential for academic and lifelong success" (Heaven 2023).

> We have recognized that, in the face of datafication and algorithmization, which obtain performance results through data interpretation, the challenge lies in environments that invite us to think "outside the box", that encourage flexibility in decision-making, autonomy and critical thinking, as well as the critical capacity to empathize in the years to come. We believe that intelligent systems that are growing in adaptive and contextualized learning environments can work in detection and recommendations, but offer few opportunities for creative and critical thinking (Lion 2022, p. 168).

7.3 Natural Intelligence Versus Artificial Intelligence

As "AI is neither artificial nor intelligent" (Crawford 2022, p. 29), it is essential, especially in the field of higher education, to understand what we mean by intelligence. In the same book, the author quotes Dreyfus, who points out that "human intelligence and expertise depend to a large extent on many unconscious and subconscious processes, while computers require all processes and data to be explicit and formalized" (Greenberger 1962, pp. 315, cited by Crawford 2022). In this section, when I mention "natural intelligence", what I am referring to is human intelligence, that "capacity of a subject to direct their behaviour, using information captured, learned, elaborated and produced by themselves" (Marina 2006, p. 9). This intelligence involves, on the other hand, several different types of intelligence, as described in the theory of multiple intelligences (Gardner 2014). Developing these intelligences is the primary task of education; to this end, it is feasible (one might almost say inevitable) in today's reality to make use of AI. The first thing that should not be done is to ban its use, as was often done with other technologies in the past, with consistently negative results. The second is to learn how to use it by reading appropriate sources, practicing its use, collaborating with teachers and, why not, together with students. In addition, it is necessary to understand that, like all technology, it is not neutral; and it can incorporate biases and inaccuracies of all kinds.

> AI is not only a part of the process of rethinking and intervening in the world, but a fundamentally political way of making the world, although it is rarely recognized as such. This process is undemocratic, dominated by the big AI houses, which are made up of half a dozen companies that dominate large-scale planetary computing (Crawford 2022, p. 43).

Indeed, the results offered by applications such as ChatGPT are based on biased data or data that is analysed with inadequate algorithms; therefore, the original and perhaps unidentified biases can become more noticeable and have a greater impact (UNESCO 2021b).

A simple and practical way to help determine in which cases it is safe to use ChatGPT is to carry out the recommended analysis in a UNESCO document (2023), which is reproduced in Fig. 7.3 (Tiulkanov 2023).

Therefore, we need to find a way for AI to be put at the service of the development of natural intelligence. In this sense, it is essential, as I detailed in Chap. 5, for education to promote critical thinking, which requires the following skills:

- Actively search for all the facets of an argument.
- Test the soundness of the statements made.
- Test the soundness of the evidence used to support the statements.

The question would be, then, how to put AI tools at the service of critical thinking as a path for the development of natural intelligence. Therefore, the dilemma posed in the title of this section is not such, since, if it were admitted that natural or human intelligence were replaced by AI, Harari's prophecy regarding the hacking of the operating system of human civilization would have been fulfilled.

This chapter is not intended to describe specific strategies for using AI in education. In fact, several very valuable guides have already been published that can be used as a reference, such as:

- UNED (Spain)
- Ministerio de Educación (Chile)
- Tecnológico de Monterrey (Mexico)
- UNESCO 1
- UNESCO 2
- UNAM (Mexico)

One of the keys to effective education is for students to learn to ask good questions, rather than just giving the right answers. "If we want students to continue to develop their curiosity, build critical thinking and reflection and research skills, we need to give space to their concerns, and also teach them to ask good questions" (Furman 2021, p. 150). And that is one of the key strategies for the appropriate use of AI in the classroom. The quality of the response offered by GAI systems depends, to a large extent, on how the question is asked, which, in technical jargon, is known as

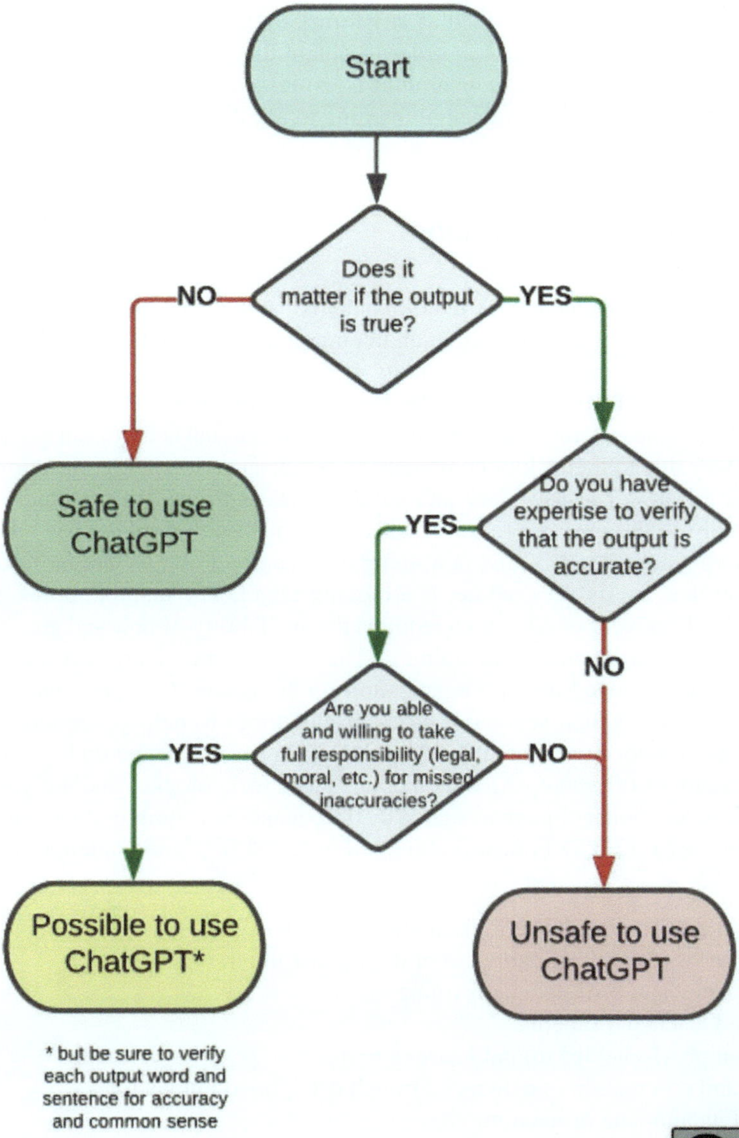

Fig. 7.3 When is it safe to use ChatGPT?

the prompt; and which can be described as any form of text, question, information or coding that is presented to the AI in order to obtain the response being sought. Depending on the way the question or prompt is expressed, the AI could provide different answers. Hence the importance of learning to ask the right questions.

7.4 Challenges and Opportunities

As was evident during the recent COVID-19 pandemic, access to digital technological resources, also known as information and communication technologies (ICT), was a decisive factor for inclusion and equality of conditions for both students and teachers. During that unfortunate period, we were able to verify how many of them were excluded from the possibility of continuing their educational activities in a relatively normal way due to the lack of computers or an adequate Internet connection, when they did not have the necessary and sufficient skills to use these technologies appropriately. Although, as a result of what I have just said, progress has been made in improving access to ICT and in training for its correct use, we are far from reaching a goal that, by its very nature, is a moving target. The tangible proof of this is, precisely, the advent of AI, which requires the availability of new and greater resources, as well as the consequent training. These issues are clearly set out in a recent document produced by the Organization for Economic Co-operation and Development (OECD), which sets out a series of guidelines to help governments, teachers' unions, teachers and other educators engage in a constructive dialogue to seize the opportunities offered by AI and other advanced technologies, and mitigate their risks to the educational objectives shared by the educational community: equity, quality and efficiency (OECD-Education International, 2023). These guidelines are as follows:

1. Equitable access to affordable, high-quality connectivity.
2. Equitable access to and equitable use of digital learning resources.
3. Teacher agency and professional learning.
4. Student and teacher wellbeing.
5. Co-creation of AI-enabled digital learning tools.
6. Research and co-creation of evidence through disciplined innovation.
7. Ethics, safety and data protection.
8. Transparency, explainability and negotiation.
9. Human support and human alternatives.

These guidelines are sufficiently developed in the aforementioned document, so it is advisable to refer to them for their correct interpretation. In particular, due to my own "professional (de)formation" as a researcher, I find what is described in Point 6 of the above list particularly significant. For this reason, at the time of writing this book, I am leading a research project on the subject within the scope of my university. In this sense, I find the conclusions offered by a recently published article very indicative of the research task, in which several fundamental avenues of research and development are presented to advance the integration of AI in education. (Walter, 2024):

1. **Curriculum Integration**: Future research should explore effective methods for integrating AI literacy across various educational levels and disciplines.
2. **Ethical AI development**: Investigating how to develop and implement AI tools that are transparent, unbiased, and respect student privacy is essential for ethical AI integration in education.
3. **AI in Policy Making**: Understanding how AI can assist in educational policy-making and administration could streamline educational processes and offer valuable insights.
4. **Cultural Shifts in Education**: Research into how educational institutions can foster a culture of critical and ethical AI use, promoting continuous learning and adaptation, is crucial.
5. **Longitudinal Studies**: There is a need for longitudinal studies to assess the long-term impact of AI integration on learning outcomes, teacher effectiveness, and student well-being. So far, this has not been possible due to the novelty of the technology.

AI is just another technology and, like any other technology, it can be used appropriately or not. The big challenge is to understand and learn how to use it for the benefit of humanity; and, as far as education is concerned, for the benefit of our students' learning.

> Just because a machine can do a job, it does not mean we should abandon it. Just because there are trains, we are not going to stop walking. We shouldn't stop thinking in numbers, even if there are calculators, or memorizing relevant data, even if it's on Google. And, in the same way, it does not seem like a good idea to do away with reasoning and the development of ideas, even if we have ChatGPT (Sigman & Bilinkis, Artificial. La nueva inteligencia y el contorno de lo humano 2023, p. 112).

Natural intelligence is what characterizes us as human beings. If this characteristic were to be left to machines, we humans would be doomed to be replaced by them.

Chapter 8
About Change

Change is the only constant.
Heraclitus of Ephesus.

You may be wondering why there is a chapter on change in this book about education … Change implies learning, and there is no learning without change. Change is, as the quote at the beginning of this chapter points out, something that happens constantly in our lives, whether we admit it or not. We change when we grow, when we learn something, when we meet someone … Life is change. However, and paradoxically, we often resist change and settle into that comfort zone in which we feel we know, or pretend to know, what is going to happen. We demand change from others, but, at the same time, we find it difficult to face the changes that depend on ourselves.

You should watch a video before continuing with the reading: https://bit.ly/Zona_de_confort

In this video, a series of questions related to change are presented (see Fig. 8.1).

Throughout this chapter, we will try to answer these questions. But, before doing so, it is good to know the phases of the emotional response to change (see Fig. 8.2). Understanding this process can help us accept that, most likely, we will have to go through these states of mind when faced with the need to make changes. This way, knowing what to expect from ourselves, we will be able to deal better with these feelings, and go through the stages more quickly and with less resistance.

Let's now try to answer the questions posed previously.

U. R. Cukierman, *The Art of Teaching, the Magic of Learning: Towards Innovative Higher Education Systems*, Studies in Systems, Decision and Control 697,
https://doi.org/10.1007/978-3-032-35021-3_8

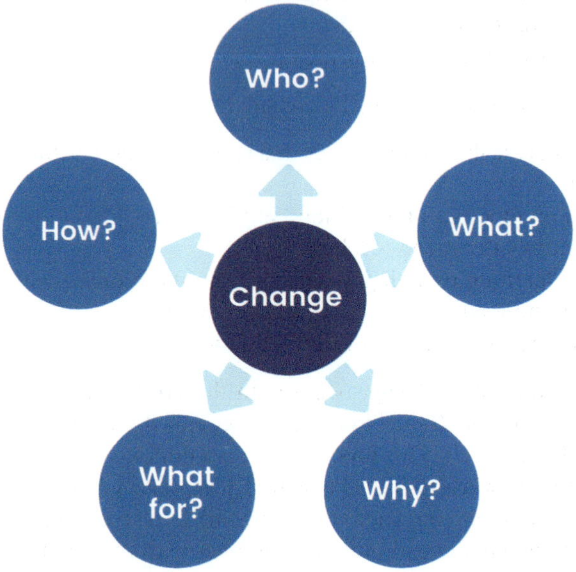

Fig. 8.1 The questions of change

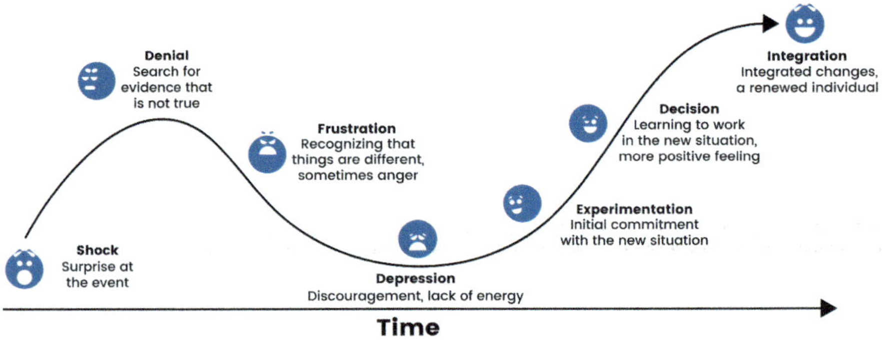

Fig. 8.2 Stages of the emotional response to change

8.1 Who?

It is common to pretend that change depends on others, and so nothing changes. This is verified daily at different levels. In the field of education, change can involve the institution, teachers and students. Let's start with the first of these levels.

8.1.1 At an Institutional Level

According to Schein (2009), there are three levels in organizational culture (see Fig. 8.3).

These three levels range from the most visible, such as organizational structures and processes, to those that are usually invisible, which the author calls "underlying assumptions": that which is not written, but is assumed to be true. Usually, one starts by changing the "artifacts", which is the easiest thing to change, and then tries to change the "proposed values"; but it is difficult to change those underlying assumptions. Real organizational change can be achieved by reversing the process; trying to change those issues that are invisible and then gradually modifying the higher levels. The culture of an organization is very stable and difficult to change because it represents the accumulated learning of its members. The important parts of the culture are essentially invisible. Resistance to change is due to the fact that the new behaviour that must be learned requires "unlearning" what we know how to do and have always done. It is natural and predictable that there will be resistance to change, since it is possible that some beliefs, attitudes, values and assumptions may have to be renounced, as well as learning some new ones.

Managing the change in organizational culture is the process by which the transition to a new reality is managed. This management seeks to convince and persuade

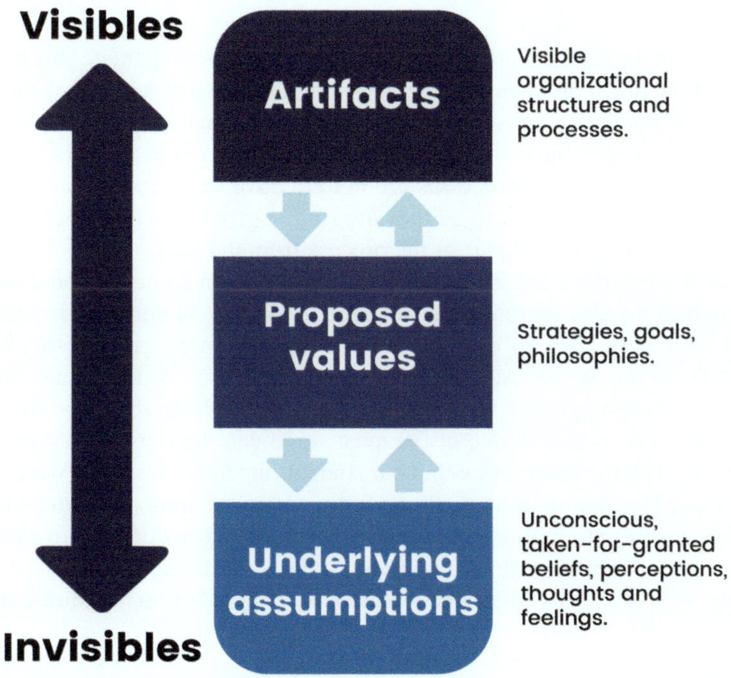

Fig. 8.3 The three levels of organizational culture

members to work in a new context that will bring good results to the organization. However, it is not so easy to change the culture and objectives of an organization, so it must be done in an organized and planned way.

One of the most cited articles on the subject is the one written by Kotter (1995), in which he proposes eight steps for managing change in organizational culture:

1. Create a sense of urgency.
2. Build a guiding coalition.
3. Form a strategic vision and initiatives.
4. Enlist a volunteer army.
5. Enable action by removing barriers.
6. Generate short-term wins.
7. Sustain acceleration.
8. Anchor the changes in corporate culture.

Whether at departmental, faculty or university level, these steps constitute a simple and clear guide on how to manage change. And it can be applied to a particular situation, such as, for example, changing the curriculum or implementing a program to improve teaching.

8.1.2 At the Teaching Level

In my opinion, we teachers are the most important agents of change. This is because, no matter how much an institution decides to undertake a process of change, it will not reach the students—the very reason for an educational institution's existence—unless we take ownership of it and put it into practice in our daily work. This is why many institutional change processes are not effective: teachers are not included as participants from the beginning, or they are simply given directives, but they are not incorporated as an essential part of the process from the start.

However, even if we are included in the process from an institutional leadership perspective, if there is no real predisposition to change on our part, it will be very difficult for the changes to crystallize and result in improvements for the whole community. In this sense, traditions and mandates tend to be significant stumbling blocks. The fairly ingrained idea that "if I was able to learn this way, my students will be able to do it too" is one of the most significant barriers to change. But this belief is based on the assumption that our students are the same as we were when we were in that place, but nothing could be further from the truth. Our students belong to another generation, with customs and expectations very different from those we had when we were in that place one, two or more decades ago. And, what is more, the world we live in is not the same as it was back then. A popular phrase describes this situation very clearly:

Fig. 8.4 Change at the teaching level

We have 19th-century educational institutions with 20th-century teachers and 21th-century students.

So let's put nostalgia aside and accept that we have to do things differently. To do that, we have to recognize the differences (see Fig. 8.4).

We must accept that we are no longer the "owners" of knowledge. Knowledge is available in many places and in very different forms; moreover, it is changing and endless, as I have described in a previous chapter. Therefore, we have to step aside from center stage and put the students there. Of course, this is neither easy nor automatic. It requires us to assume a new role, which is not less important than the previous one, but different: a role of guide, of facilitator. Those who are parents of adult children know this change of roles. When our children are young, we are the ones who have all the answers for them, the ones who know everything there is to know. But, as they grow up, our place and our role change: we become the ones who accompany them in their own discovery of the world, who encourage them towards their own construction of meanings. In short, it is a matter of reframing our role as teachers in order to continue being a relevant actor in this new scenario in which we currently find ourselves. To paraphrase the great Mahatma Gandhi, teachers must be the change we want to see in our institutions.

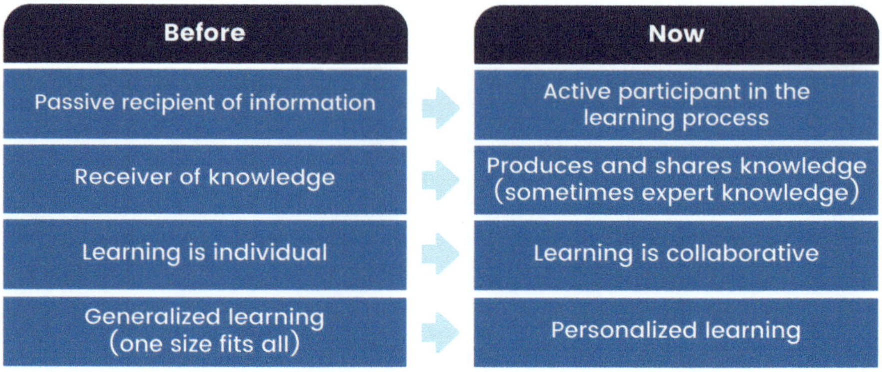

Fig. 8.5 Change at student level

8.1.3 At a Student Level

Historically, students were seen as the 'recipients' of knowledge that teachers 'transmitted' to them (unfortunately, in many cases, this is still the case today). This terminology, more appropriate for a communications system than for the teacher-student relationship, is still in force in many educational institutions. In this scenario, the student is a passive subject who receives knowledge. Changing this reality requires students to be active participants in the change. Also, in this case, we can list what these changes are about (see Fig. 8.5).

I can almost "hear" the objections that many colleagues might be making to this approach: that students want everything spoon-fed, that they are not interested in anything, that they are unmotivated; that what they are most concerned about is how to pass the course, not so much how they are going to learn what they need to know. Experience has shown me over many years that, with the right encouragement and support, this is the least of the problems. Teachers and institutions are much more resistant to change; in fact, their young age makes them much more malleable and adaptable to circumstances. My proposal, then, is to make them "partners" in our initiatives for change, for which it will be essential that they understand why and for what purpose we are doing what we are doing; and that, even when it requires an extra effort, they will be the main beneficiaries of the change.

8.2 The What, the Why and What for?

In the workshops for teachers that I give, I usually ask the attendees what needs to change. The answers tend to vary: the curriculum, the methodology, the content, etc. Some even answer "everything". But, before moving on with the subject, I suggest you stop reading and watch a very short video.

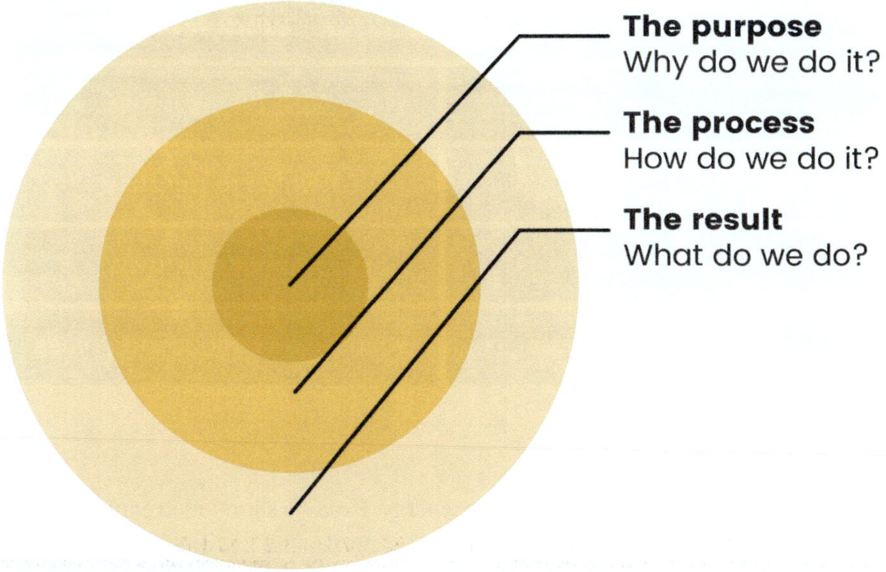

Fig. 8.6 The golden circle

The Golden circle: https://bit.ly/el_circulo_dorado

In Fig. 8.6, I describe what Simon Sinek explains in the video.

Let's try to adapt the logic described by Sinek to our reality. We can easily answer what we do; almost certainly, we can also answer how we do it; but perhaps answering why we do it conscientiously is a little more difficult. The truth is that it would be much better and more effective to start by answering the question of why we do what we do, and then answer the other questions.

When we talk about change, it is important to define the what for instead of the why. In Fig. 8.7, the reasons that justify what I am saying can be clearly seen.

When asked why change, reactions such as "I do not see the need to change, so why bother" or "they are forcing me to change, so I'm going to resist" may arise. Change is not easy, but it is necessary. Even those who believe that everything is fine and that, therefore, nothing has to change will soon notice that everything around them has changed and, consequently, they will have to change. The important thing is to have a purpose, a reason why, that gives meaning to the change we are going to make, whether on a personal or institutional level. As stated in the book Change the chip, "many times, what seems like resistance is a lack of clarity" (Heath and Heath 2018); In other words, the purpose has yet to be identified. In the same book, the authors refer to an analogy, originally proposed by University of Virginia psychologist Jonathan Haidt, who says that our emotional side is an elephant and our rational side is a rider. If we want things to change, we have to keep both in mind: the rider provides

Why?	What for?
Seeks the causes (the past)	Aims for the future
Focuses on the reasons that brought us here	Relates to purpose
Easy to answer	Requires creativity
Focuses on behaviors	Focuses on beliefs and values

Fig. 8.7 The why and the what for

planning and direction; and the elephant, energy. Here's a short video explaining this analogy.

The rider and the elephant: https://bit.ly/jinete_y_elefante

As Jonathan Haidt puts it, "if you want to change people's minds, you have to talk to their elephants", that is, to their emotions; give change a sense, a meaning, a purpose. That goes for others and, very especially, for ourselves.

When it comes to promoting change at an institutional level, leadership is required to drive it. In the report produced by the National College for School Leadership (NCSL 2003), leadership is defined as follows:

- Leadership is a **process of influence** that leads to the achievement of desired purposes.
- Successful leaders **develop a vision** for their schools based on their personal and professional values [...] **and influence** their staff and other stakeholders to share the vision.
- The philosophy, structures and activities of the school are **geared towards the achievement of this shared vision.**

This leadership can be exercised both at the highest level of university management and at any of the intermediate levels, such as a faculty, a department, a chair; or even at the course level; understanding that the teacher is the leader of that course and the students are the ones who should be able to share the teacher's vision, which aims for them to learn more and better.

Tim Knoster presented a model for managing complex change, which has several components, at a TASH Association conference. And he suggested that when the components of vision, skills, incentives, resources and action plan are collectively inherent in the system, change is likely to occur. However, if any of the components are missing, the "change process" can be inhibited or fail. Knoster's model has been widely adopted in the field of education to help teachers and administrators implement effective teaching practices. However, its principles are also relevant for companies and organizations seeking to embrace change. By proactively identifying and addressing the essential elements of the model, leaders can facilitate a smoother and more successful transition (see Fig. 8.8).

In a study carried out by Ruth Graham, with the sponsorship of The Royal Academy of Engineering and the Massachusetts Institute of Technology (Graham 2012), the author identifies four common characteristics for successful and widespread change that appear to be largely independent of geographical location or type of institution. Although the study focused on engineering education, the conclusions are valid for other disciplines. These characteristics are as follows:

1. A widespread recognition that **change in training is inevitable as an interaction between colleagues and common sense among teachers for the achievement of a comprehensive reform of the curriculum**.
2. Success seems to be associated with the scale of change embedded in an **interconnected and coherent curriculum structure**.
3. **The Department seems to be the engine of change**, with the relentless commitment of the Head of Department as a decisive component for its success.
4. The study highlights the **challenges associated with sustainable change**, with

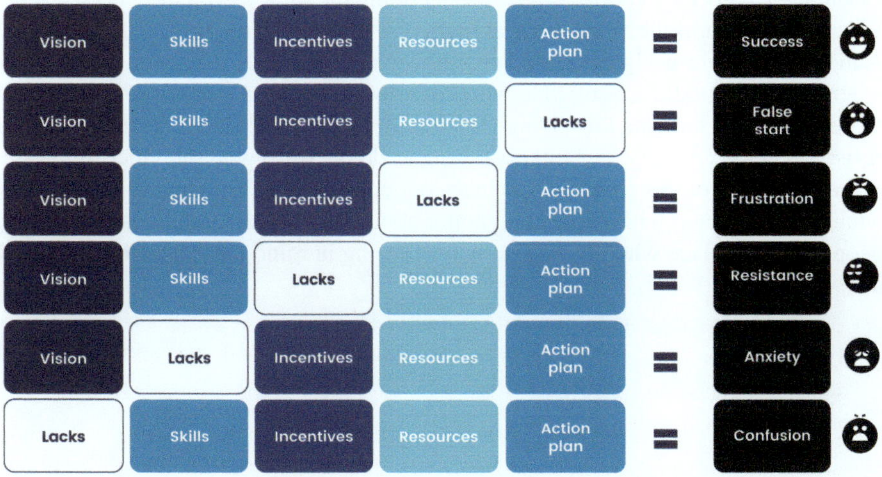

Fig. 8.8 Change management model (Adapted from Tim Knoster (1991). Presentation at the TASH conference, Washington, D.C.)

most reform initiatives reverting to the status quo in the years following their implementation.

Having reached this point, I suggest you carry out the following activity:

Activity 8.1: This activity is called "The 4 Ps". It is an exercise that will help you to face changes both on a personal level and in the institution in which you carry out your tasks.

The 4 Ps

1. **Project**—The first step is to list the upcoming change(s).
2. **Purpose**—Next, it will be necessary to analyze what benefits these changes will bring. Will retention improve? Will student outcomes improve? Etc.
3. **Parts/Requirements**—After the two previous steps, it will be necessary to establish what will be required to effectively implement the changes. For example: if the project consists of implementing student-centered learning, the requirement could be teacher training.
4. **People**—Finally, it will be necessary to identify who in the institution will need to change the way they do things or actively participate in the change.

To finish this chapter, I quote a well-known phrase, mistakenly attributed to Einstein: "Do not expect different results by doing the same as you have been doing". We could paraphrase it by saying that there can be no improvement without change. And, since we are using phrases, I remember one that a teacher used to say to us when I was in middle school: "He who does not go forward goes back". Indeed, all these statements indicate that change is inherent in life; and, therefore, it is useless to resist it. On the contrary, the smartest thing to do, as we saw in the video suggested at the beginning of this chapter, is to leave our comfort zone and dare to dream about what we understand will help us feel better about what we do, while also improving people and the place where we are, always bearing in mind that **change starts with ourselves**.

Chapter 9
About the Challenges

We need diversity of thought in the world to face new challenges.
Tim Berners-Lee

In my doctoral thesis, mentioned previously, I identified six challenges that affect teachers, institutions and knowledge in a cross-cutting way. In that work, I came to the conclusion that higher education institutions face, even before the pandemic, a series of challenges that question the status quo that has been in force for centuries, in such a way that the current education system is strongly influenced, if not completely modelled, by the realities and needs of a world that no longer exists (Cukierman and Recabarren 2018). I will now go on to develop some of these challenges.

9.1 Challenge 1: The Democratization of Higher Education

In 2000, the estimated total number of people in post-secondary education (essentially university) was approximately 443 million. By 2020, that number had doubled to 932 million, and it is estimated that by 2040, it will reach 1.56 billion students (Oxford Martin School. University of Oxford 2018). Fig. 9.1 shows these projections. Obviously, the democratization of higher education is a positive development, but it represents a challenge in terms of the demand for greater resources for higher education institutions, since it requires investments in infrastructure and human resources that not all institutions are in a position to afford, whether the economic resources are provided by the students themselves, by third parties, by the state or by a combination of these.

On the other hand, there is a risk that this process of growth in higher education enrolments will not really democratize as it does not fulfil the requirement of equal opportunities. Indeed, as clearly stated in a recent UNESCO document (2021a), "looking ahead, more should be done in fulfilling the right to higher education for al". In this sense, inclusion also means that "Higher education should be organized in a way which focuses on integrating learning across disciplines. As learners change, so too will the role of those who support them". In the same document, it is stated:

U. R. Cukierman, *The Art of Teaching, the Magic of Learning: Towards Innovative Higher Education Systems*, Studies in Systems, Decision and Control 697,
https://doi.org/10.1007/978-3-032-35021-3_9

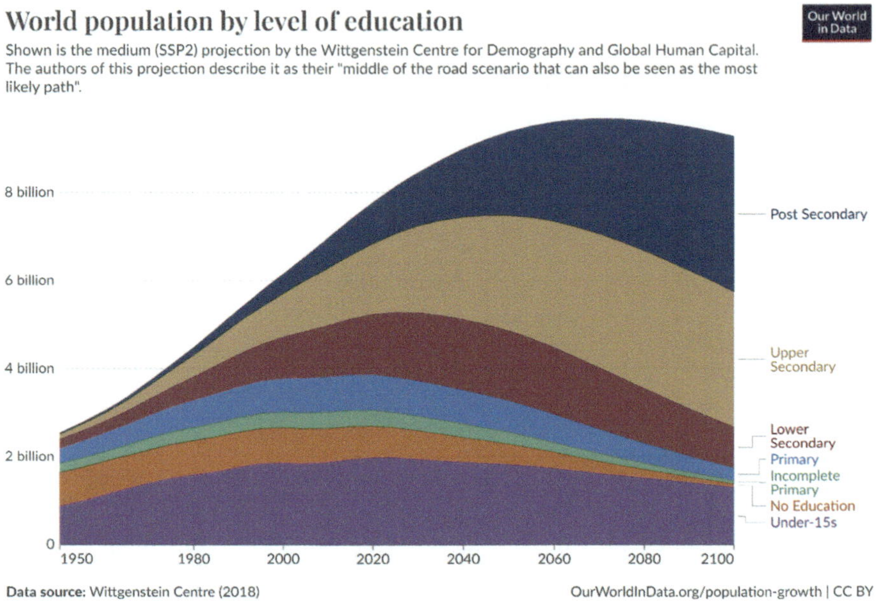

Fig. 9.1 Population with higher education

"It follows that greater focus should be placed on crafting flexible and inclusive programmes and pathways, as well as adaptive and creative approaches to assessing and accrediting learning outcomes".

This issue is directly related to the concept of "personalization of learning", which I mentioned previously.

9.2 Challenge 2: Updating Labour Demand

Higher education institutions, as the training ground for the professionals that society demands, face the challenge of adapting their training courses to these requirements. There are two main reasons for this: on the one hand, the students themselves, who pursue their university studies with the aim of obtaining a degree that will enable them to perform a professional role effectively; on the other hand, the society that sustains and is served by these institutions, which train professionals to meet the needs of society (Dondi y otros 2021). This is particularly true in the case of engineering degrees and at the present time, when there is an unmet demand for engineers (Fernández 2020; Royal Academy of Engineering 2020). The report on engineering for sustainable development by the UNESCO International Institute for Higher Education in Latin America and the Caribbean (IESALC) (2021c) describes the following:

> While engineers are crucial to promoting the Sustainable Development Goals (SDGs) and meeting the aspirations of developing nations, the world is currently experiencing a shortage in both the number of engineers and the caliber of engineering skills available (p. 21).

The situation in Argentina is no different to the rest of the world. In 2012, the national government set itself the goal of having one engineer for every 4000 inhabitants and around 10,000 graduates per year (Ministerio de Educación 2020). According to the latest available records, the number of engineering graduates per year has stagnated at around 8000 (Secretaría de Políticas Universitarias). But, beyond the numbers, the skills required of graduates are also important. In this sense, in recent decades, there has been a greater emphasis worldwide on so-called "soft skills" or "emotional competencies", which are those related to certain issues, such as effective communication, collaboration, critical thinking, creativity, teamwork, entrepreneurial capacity, etc. Fig. 9.2 shows these skills according to the results of a survey of 18,000 employers in 15 countries (Dondi et al. 2021).

In Argentina, this situation was clearly evident in the wording of the social, political and attitudinal competencies included in the graduation competencies for Engineering degrees (CONFEDI, 2018), and validated in the resolutions that define the new standards for the accreditation of these degrees, namely:

- work effectively in teams;
- communicate effectively;
- Act ethically, with professional responsibility and social commitment, considering the economic, social and environmental impact of the activity in the local and global context;
- Learn continuously and autonomously;
- Act with an entrepreneurial spirit.

The current deficit in training for the development of these skills urgently demands to be addressed. However, according to a recent survey, employers in Argentina will mainly seek out this training outside higher education institutions; only one in three companies will do so in these institutions (17.5%, in state-run institutions; and 15.9%, in private institutions) (World Economic Forum, 2020; Vilchis, 2024) (see Fig. 9.2).

9.3 Challenge 3: Automation

The first thing to do is to develop the concept of automation, particularly in relation to higher education. The idea is to analyse this concept insofar as it implies a conversion from manual to automatic processes, and also how this impacts on the training of professionals. It is important, however, to point out that we are talking about automation that involves the use of intelligent machines so that manufacturing processes can be carried out with minimal human intervention. This process has an impact on the labour force, which has been analysed and discussed at length, but on which there

Fig. 9.2 Fundamental skills and attitudes that will help citizens thrive at work

are no unanimous conclusions. On the one hand, there are those who foresee a direct effect in the loss of a significant percentage of jobs (Frey y otros 2016); and, on the other hand, those who believe that job losses due to the automation of industrial processes are reversed by indirect gains in client industries and induced increases in aggregate demand (Autor and Salomons 2018). In any case, it is clear that occupations involving routine tasks will be easier to replace with intelligent machines or robots than those requiring higher-order skills and essentially human competencies, such as those described in the previous section on the updating of labour demand. In this sense, what is required is what Aoun (2017) calls "robot-proof education". The author proposes the development of new literacies in the field of higher education, namely:

- data literacy: the ability to read, understand, create and communicate data with added value;

- technological literacy: the ability to understand, use, apply, evaluate and create new technologies;
- human literacy: the ability to interact with others and evaluate the social, ethical and existential impact.

As Oppenheimer puts it (2020):

> Governments need to get their act together, invest more in improving the quality of education, in retraining people, and in science and technology, in order to generate more sophisticated jobs that robots can hardly replace in the short term.

Furman expresses himself in a similar vein (2021) when he assures:

> We must incorporate the issue of work automation into our discussion of what is worth learning for children and young people today. Because, although we have no certainties, we do know that certain capacities will probably always remain relevant and difficult to replace, such as critical thinking, creativity, the ability to work with others and empathy. And our educational institutions, from kindergarten to university, have to contribute to training for that future (p. 170).

9.4 Challenge 4: Digital Transformation

I have already defined this concept in previous pages, but it is worth carrying out a more detailed analysis here in terms of the challenge for higher education institutions. In this sense, the result of a recent piece of research is very illustrative, from which the following fundamental findings emerge (Brooks and McCormack 2020):

- Digital transformation cannot be ignored.
- Few say their institution is committed to current digital transformation, but many are preparing to do so.
- Optimism about realizing a digital transformation in the near future is high.
- Perceptions of who knows, understands, and supports digital transformation vary greatly by campus role.
- Digital transformation is a student-centered effort.
- The impact of digital transformation on institutional functions has varied widely until now.
- The biggest barriers to digital transformation are what one might expect: cultural change and cost.

Although these conclusions were obtained on the basis of research carried out in the United States of America, many of them could be transferred unchanged to the Argentinean context.

9.5 Challenge 5: Accreditation

The Higher Education Act No. 24521 establishes the compulsory accreditation of postgraduate and undergraduate courses with "degrees corresponding to professions regulated by the State, the exercise of which could compromise the public interest by directly endangering the health, safety, rights, property or education of the population" (Congreso de la Nación Argentina 1995). Since the beginning of this century, and without interruption, these accreditation processes have been developed by the National Commission for University Assessment and Accreditation. Recently, new standards for the accreditation of engineering degrees were approved, replacing the previous (CONEAU 2021), based on the competency-based approach.

Undertaking an accreditation process (a task that is carried out, depending on the case, every three or six years) involves a significant challenge, as it involves mobilizing the entire educational community so that the assessments to which higher education institutions are subjected are satisfactory and allow the normal continuity of the courses in progress. In the current circumstances, the challenge is increased due to the emergence of a new standard and, even more so, a conceptual approach substantially different from the previous one. In effect, the previous standards followed the guidelines of a content-based approach, while the new ones that are emerging are competency-based, which fundamentally implies a change in methodologies, since they transform learning into active and student-centered learning (Cukierman 2018).

It is worth mentioning what is expected to be a new higher education law, which is currently being debated in Argentina (2021), which is a new challenge, that will be revealed when its development is finalized and it is enacted.

9.6 Challenge 6: The Pandemic

The pandemic has had immeasurable effects on all aspects of life, all over the world. Due to the characteristics of this book and my personal experience, I will only analyse its impact on higher education. Throughout the health emergency, some lessons were learned. One thing that seems clear is that digital technologies have definitively conquered the educational universe; in particular, in higher education. As Rivas says (2021):

> This forced migration to the digital educational cloud has caused a thaw. Suddenly, it shows us the forms that education can take in increasingly technology-driven societies. It is not only the digital consumption of educational screens and algorithms that is changing. There are also deeper transformations: the curriculum is expanding and diversifying, new avenues for autonomous learning are opening up, pedagogies are being redesigned in more diverse, fragile and dynamic cultural environments (p. 5).

To complete this section dedicated to the challenge generated by the pandemic, it is appropriate to quote Rivas (2021) again, who, a few pages later in the same article, says:

> The biggest challenge is to generate ecosystems for the development of quality educational platforms that integrate face-to-face and digital education, that respect and engage in dialogue with teachers, and that seek to guarantee the right to education. The pandemic has changed everything and has opened up scenarios for rethinking education when things return to a certain degree of normality. Hybrid models are an opportunity to redefine the meanings and purposes of education in a changing and unequal society. This conversation has just begun (p. 17).

However, we should not see this situation of accelerated and massive adoption of digital technologies as something inexorable. In any case, we should see it as an opportunity to review educational practices and generate the necessary reflections to avoid deepening situations of inequality, which also became evident during the emergency. In this sense, it is relevant to quote Dussel et al. (2020):

> More complex approaches indicate the need for a different way of relating social and institutional actors, the State and technologies, taking into account the deployment of the actors' own subjectivity, selective and strategic interventions by the State, with necessary investments in the technological field, responding to concrete and situated demands. This option involves technologies that must be developed and appropriated for their own uses. This perspective implies an invention of relevant content for teaching and learning, the development of public platforms and the deployment of collaborative portals with shared resources (p. 364).

9.7 To Conclude on the Challenges Posed

So far, I have described the six challenges facing higher education today, namely: democratization, keeping up with labour demand, automation, digital transformation, accreditation and the pandemic. The combination of these six challenges places higher education in an unprecedented situation, requiring concrete actions to remain relevant in and for the society it serves. But it would be a mistake to believe that these challenges are new or that they have arisen with the pandemic. Indeed, as early as 2018, more than 600 university rectors from 26 countries proposed a number of important actions and programs:

> The debates that arose in Salamanca 2018 point to certain actions and programs that are especially relevant for universities, such as: making innovative educational methods more flexible and applying them and rethinking organizational, administrative and sustainability processes; alliances, courses and certifications with companies from different industries; new and alternative models of certification and integration with global platforms; hybrid training offers and workplace training and refresher programs, in the context of training that is adapted to the needs of the student and that extends throughout life; new degrees, especially those related to computer science, artificial intelligence, data science and technology; and a greater emphasis on humanistic education, as well as on the transversal competences of students (Declaración de Salamanca, 2018, p. 1).

Another essential reference for this analysis is that proposed by the United Nations Organization (2015) through the sustainable development goals. One of these goals, the fourth, states that there should be "inclusive, equitable and quality education and promote lifelong learning opportunities for all". In this sense and in line with this objective, in May 2015, in the city of Incheon (Republic of Korea), more than 1600 participants from 160 countries met and issued a declaration (UNESCO 2016) which focuses on three fundamental axes: quality, inclusion and innovation. With regard to the first axis, it is necessary to "promote quality lifelong learning opportunities for all, in all contexts and at all educational levels" (p. 7). Likewise, the document states that "inclusion and equality in and through education are the cornerstone of a transformative education agenda" (p. 7). Finally, in terms of innovation, it states that "it is essential to take advantage of innovation and ICTs to strengthen education systems" (p. 32).

Bearing in mind what has been said so far, we can summarize the challenges listed above in terms of three axes:

(a) Increasing the quality of higher education means improving learning and reducing dropout rates.
(b) Promoting inclusion means designing inclusive pedagogical practices in the broadest sense.
(c) Innovation can be boosted by taking advantage of emerging technologies

Various studies demonstrate the benefits of using and combining emerging technologies to improve and transform different educational processes (Agreda Montoro et al., 2019; Martin et al., 2020; Sosa Neira et al., 2017).

However, it is important to point out that it is "necessary to improve the coordination between the challenges presented by the technological landscape and how society is educated to face these challenges. Poor coordination between these two worlds could present serious challenges" (Cobo 2019, p. 29).

9.7.1 Conclusions on Virtuality in Higher Education

As I have already said, as a result of the pandemic, the use of technological resources in education in general, and in higher education in particular, accelerated and was boosted in a way that would have been unthinkable a short time before. Thus, the pre-pandemic situation, in which university education was mostly face-to-face, migrated completely and forcibly to virtuality. With the continuation of the pandemic, the benefits of virtual education became evident. Now that we have overcome the pandemic, there is a strong trend towards the use of hybrid models, understood as those that combine face-to-face and virtual learning in ways that depend on each case and situation. In this sense, I recommend reading the book *Hybrid: Teaching at the university that we did not see coming* (Maggio 2023). It is worth mentioning at this point that there is a continuum between presence and absolute virtuality, as represented in Fig. 9.3.

However, we should not underestimate certain risks in virtuality:

- technological disconnection, due to a lack of connectivity in homes;

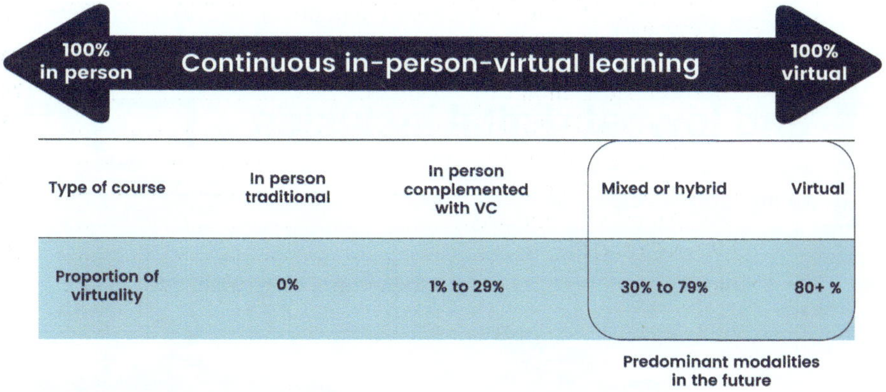

Fig. 9.3 Hybrid modality. *Note* Adapted from Allen et al. (2007)

- personal disconnection, due to the low capacity for self-regulation of learning and the socio-emotional drifts that students present;
- pedagogical disconnection, due to an inappropriate pedagogical proposal for the channel and the context.[1]

Figure 9.4 shows some data referring to digital inclusion, where it can be seen that, even in the best positioned countries in the region, there is still a significant difference with respect to the most advanced countries in this area.

On the other hand, we must point out that face-to-face learning also has its own problems, namely:

- emphasis on master classes;
- little interaction between students and with the teacher;
- very large classes;
- higher costs;
- more rigid schedules.

Personally, I believe that the hybrid modality should be an opportunity to take advantage of each one; in this sense, below I propose what I have come to call “meaningful presence”.

9.7.2 *The Concept of Meaningful Presence*

In the 1960s, the concept of meaningful learning (Ausubel 1960, 1963) was introduced, often contrasted with rote learning, associated with traditional expository classes and summative assessments. What I propose, then, is that the activities carried

[1] According to Francesc Pedró, director of UNESCO IESALC, September 2021.

Fig. 9.4 Digital inclusion. *Note* Taken from Chevalier Naranjo (2021)

out in face-to-face meetings produce meaningful learning, that they are not activities that the student can easily do on their own at the time and place that is most appropriate for them, such as, for example, readings, videos, research, etc. Maggio expresses this wisely (2021): "My aspiration is that each meeting has a life of its own, that it is a moment in which we all want and need to participate… A meeting should be a **lively and vivacious experience**" (pp. 129–30). The aim, therefore, is to promote active, student-centered learning in face-to-face settings (Cukierman 2018), and to take advantage, as much as possible, of the flipped learning methodology, which I have described in detail in Chap. 2 of this book. Likewise, the aim is to take advantage of the benefits of collaboration tools and digital platforms in virtual reality, and to seek to personalize learning through artificial intelligence tools.

9.8 Proposals for Facing these Challenges

Throughout this chapter, I have proposed not only to understand reality, but also to propose scenarios and ideas that can lead to the transformation of that reality. Therefore, based on what has been said so far, I will briefly propose some actions necessary to face the challenges that teaching presents in the immediate and medium-term future in higher education institutions.

The first step is to outline interdisciplinary collaborative strategies. To do this, alliances and agreements must be established between higher education institutions that enable students to work together and collaborate with peers from different disciplines. The reality of any professional's work shows that the problems they face cannot be solved individually, but require the intervention of professionals from different disciplines. In this sense, technology opens up a wide range of possibilities based on the proliferation and appropriation, by teachers and students, of digital collaboration tools. This makes it possible to think of interdisciplinary activities both locally and internationally, as well as to acquire essential skills for professional work whenever daily work practice requires the efficient and effective mastery of these collaboration tools.

The second point is that it is necessary to promote educational innovation. As I have already said, innovation does not necessarily involve the inclusion of technology in educational activities. Good proof of this can be found in the aforementioned practice of some teachers who, forced to use videoconferencing tools during the pandemic, transferred lectures from the face-to-face classroom to the virtual classroom without changing anything in the process. In this sense, it is relevant to quote Rivas (2017), when he proposes his definition of educational innovation:

> With this horizon setting the course, we will define educational innovation as a vital force, present in schools, educators, projects and policies, which is capable of recognizing the limitations of the traditional educational matrix and altering it for the benefit of our students' 21st century learning rights. In other words, to innovate is to alter the elements of a school order that extinguish or limit students' desire to learn (p. 20).

Promoting educational innovation involves working on various aspects, such as teacher training or the promotion of research in the field with a perspective specific to each discipline; and it also involves generating incentives for those teachers who transfer innovations to their classrooms. These incentives can be financial, although this is sometimes difficult or impossible in state-run universities, as salaries are defined by other parameters; or in the form of explicit recognition from the authorities, peers and students; or through surveys that are routinely conducted.

The third point is the need to incorporate emerging technologies. In various parts of this book I have presented and analysed emerging technologies which, according to a variety of sources, show the greatest potential in the short and medium term for use in higher education. Regarding this, it is important to take a critical stance to avoid pressure from large companies leading to the use of these technologies in ways that are less suitable from a teaching point of view; and, even worse, to avoid widening the gap between higher education institutions in developed and developing countries. The following is worth quoting here, as expressed in a recent CELAC report (2021):

> The technological revolution, together with the change in the strategies of the leading companies in the use of digital technologies, has led to the rise of global platforms, giving rise to an excessive concentration of economic and political power in no more than twenty corporations of two or three world powers, an excessively small group of companies whose market value exceeds or approaches a trillion dollars (p. 7).

To this end, the implementation of research and development projects for educational technologies in line with specific needs should be encouraged, using the criteria of free and open source software. It is worth highlighting the existence of technological and social development projects that equate this type of development with traditional scientific research projects, thus enabling these activities to be considered for the categorization of those involved.

The fourth point is the need to train teachers. Unfortunately, the majority of university teachers have little or no pedagogical training. It is strange to see that, in order to teach at university, the only requirement is training in the subject to be taught, but not training in the teaching profession. I do not intend in this text to restrict teaching activity in higher education institutions, as it is currently conceived, but what is necessary is that, as is the case in each of the professions that are practiced outside the academic sphere, a plan for professional updating be proposed and developed in those universities that are not yet doing so, a plan that will provide all teachers with the opportunity to achieve the necessary suitability to practice this other profession with at least the same suitability with which they practice their basic profession. In fact, in some universities, such as the UTN, where I have developed my teaching career for almost 40 years, this is a requirement of the academic career, which establishes in article 1 of the corresponding Ordinance that "teaching improvement will be continuous and institutional" (Consejo Superior UTN 2008). To this end, it would be advisable to have specific funds allocated by the relevant authorities.

The fifth point is that it is necessary to update the modalities. As I have described, the pandemic led to a revaluation of distance learning. Although its origins date back to the mid-nineteenth century, when Isaac Pitman introduced correspondence (Tait 2003), the greatest development of this methodology occurred in the mid-twentieth century; in particular, with the spread of the Internet and digital communication and collaboration tools. Thus, it is important to highlight the democratizing intention that the modality had in Argentina since the creation of the Argentine University Network for Distance Education (RUEDA) and through its promoter and first coordinator, the remembered Edith Litwin. Indeed, as Coicaud explains (2002), another of the long-standing coordinators, the modality encourages the role of "educators committed to the democratic purpose of providing people with equal opportunities to learn by breaking down physical, psychological and social distances, defending the public nature of knowledge" (p. 11). The author also points out that:

> The historical development of distance education in different times and places is not subject to the mere evolution of technology, but to its potential to promote genuine learning in people, since [...] the use of the best means to study in this modality is more related to the

> implementation of good teaching and learning proposals than to the intrinsic characteristics of the same (Coicaud 2002, p. 10).

As I mentioned previously, today we need to avoid restricting modalities based on normative or instrumental criteria, and instead be able to select scenarios based on pedagogical needs with the help of technological tools in order to promote and facilitate access to higher education for more students with equal opportunities, regardless of time and space; whenever possible, taking the possibilities and limitations of institutions, teachers and students, and the available tools into account. It is also here, with particular reference to state-run higher education institutions, where the role of the state is essential.

The sixth point is that it is essential to guarantee equal access to the technological tools necessary for the development of teaching and learning activities. In line with what I have stated in the previous paragraph, there is no doubt that the State must guarantee the achievement of this objective. As expressed in the Incheon Declaration, "education is a public good, the main guarantor of which is the State" (UNESCO 2016, p. 28). At the national level, the Digital Argentina Act is applicable, which, in its first article, states:

> The development of Information and Communication Technologies, Telecommunications, and their associated resources is declared to be in the public interest, establishing and guaranteeing the complete neutrality of the networks. Its aim is to enable access for all inhabitants of the Argentine Republic to information and communication services under equitable social and geographical conditions, with the highest quality parameters (Congreso de la Nación Argentina 2014).

Argentine governments have developed different plans in this regard with varying degrees of effectiveness, an issue that is not analysed here. I can mention the "argentin@internet.todos" Program, the National Program for the Information Society, the Educ.ar Program, the National Telecommunications Plan "Argentina Conectada", the "Conectar Igualdad" Program and, more recently, DNU 690/2020, through which the national government ensures connectivity for the entire population (PEN 2020). However, achieving the objective set out here requires the participation of other actors, including communication service providers and computer equipment providers, the vast majority of whom are currently private. This is where universities, and in particular faculties offering engineering or ICT-related degrees, can actively collaborate. As proposed by Sábato and Botana, through the famous Sábato Triangle, which promotes interaction between the state, the scientific-technological infrastructure and the productive sector (Sábato 1979). I would like to mention the role that the UTN, the university where I graduated as an engineer and where I have been working for more than 40 years, can play in this matter, as it has a long history of varied, diverse and fruitful examples, due to its territorial dispersion and disciplinary specificity. Therefore, it is about promoting these actions to achieve the aforementioned objective for the benefit of the institution itself and society in general.

Finally, training pathways need to be made more flexible. As I mentioned previously (Declaration of Salamanca 2018; UNESCO 2016), this flexibility would make it possible to promote more personalised learning, while also enabling intermediate qualifications. Several actions have already been developed in this regard, such as the National System of Academic Recognition (Ministerio de Educación y Deportes 2016) and, more recently, the document drawn up by the Ministry of Education, the Secretariat of University Policies and the rectors of state-run universities. Among other objectives, this document proposes a review of the difference between the theoretical duration and the actual duration of degree courses; and the creation of intermediate qualifications, certifications and the recognition of skills; for both initiatives, a more flexible approach to curriculum design is required (UNS 2021). Regarding curricular flexibility:

> Conceptually, it refers to the form that a curriculum takes in order to allow its adaptation and transformation in accordance with scientific, technological, professional and social changes, on the one hand, and the construction by the student of his or her own curricular itinerary, on the other. It implies greater freedom for the student, possibilities of choice and intra- and inter-institutional mobility. In short, it is a form of curricular organization that delegates part of the curricular decisions to the students and, in turn, allows for the incorporation of new scientific or professional developments (UNRC 2017, p. 10).

In the specific case of the UTN, the Higher Council approved the general guidelines for curriculum design for Engineering degrees, which state:

> A flexible Curriculum Design should establish the basic contents in relation to the specific competences, allowing for their deepening according to the requirements of each region, the projects of each Regional Faculty, the social commitment and the needs for updating (Consejo Superior UTN 2020).

Throughout this chapter, I have described six challenges that lead to the proposal of seven necessary actions to face the immediate and medium-term future of higher education institutions. Below, I list a series of actions necessary to face these challenges:

1. Establish interdisciplinary collaborative strategies.
2. Promote educational innovation.
3. Incorporate emerging technologies.
4. Train teachers.
5. Update modalities.
6. Guarantee equal access to the technological tools necessary for the development of teaching and learning activities.
7. Make training pathways more flexible.

These actions are not static or immutable. On the contrary, they require constant review and updating according to the situations that arise and the actors involved. It is therefore necessary to conduct research on practice in the workplace and focus on improving aspects of teaching practices themselves (Blaxter y otros 2010).

9.9 One Final Thought

Below I reproduce a story that, according to my research, has been adapted from a Sufi.[2]

Once upon a time there was a writer who lived in a house on the beach, next to a small fishing village. Every morning, he would take a walk along the seashore to find inspiration and he would spend his afternoons writing at home.

One day, while walking along the beach, he saw a lump that seemed to be dancing. When he got closer, he saw that it was a young man who was amused by picking up starfish from the sand, one by one, to then throw them into the ocean.

"Why are you doing it? "the writer asked him.

"Can't you see?" the young man replied with ease." The tide has gone out, and the sun is really hot. If they stay in the sand, the sun will dry them out and they will die."

"But, man, there are thousands of kilometers of coastline around the world and hundreds of thousands of starfish scattered along its beaches. What good will it do to throw a few of them back into the ocean? Most of them will die anyway."

The young man picked up another starfish from the sand and threw it into the water; then he looked at the writer and said:

"It'll be good for this one at least."

That night, he couldn't sleep; he couldn't even write. Early the next morning, he went to the beach. There he met the young man, and together they began to collect starfish and throw them into the ocean.

My dear colleague, I invite you, as you have honoured me by reading this text, to accompany me as I collect starfish and throw them back into the ocean. Sometimes, in our teaching work, we feel like the writer in front of the young person. But let us be certain that, if through our actions we have been able to achieve at least a small change that helps someone advance on their learning path, our effort will have been worthwhile.

[2] Sufism is the internal dimension and the spiritual aspect of Islam (Source: Wikipedia).

References

1. Agreda Montoro, M., Ortiz Colón, A., Rodríguez Moreno, J., Steffens, K.: Emerging technologies. Analysis and current perspectives. Digital Educ. Rev. (35), 186–210 (2019). Retrieved enero 10, 2022 from https://revistes.ub.edu/index.php/der/article/view/27395/pdf
2. Algreen-Ussing, H., Fruensgaard, N.O.: Metode i projektarbejde. Aalborg University Press, Aalborg (1990)
3. Alhazmi, A.K., Rahman, A.A.: Why LMS failed to support student learning in higher education institutions. In: IEEE Symposium on E-Learning, E-Management and E-Services, pp. 1–5. Kuala Lumpur: IEEE (2012). https://doi.org/10.1109/IS3e.2012.6414943
4. Allen, I.E., Seaman, J., Garrett, R.: Blending In. The Extent and Promise of Blended Education in the United States. Sloan-C, Boston (2007)
5. Alliaud, A.:Los artesanos de la enseñanza. Acerca de la formación de maestros con oficio.Paidós, Buenos Aires (2017)
6. Amare, N.: To slideware or not to slideware: students' experiences with powerpoint vs. lecture. J. Technical Writing Commun. **36**(3), 297–308 (2016). https://doi.org/10.2190/03GX-F1HW-VW5M-7DAR
7. Ambrose, S.A., Bridges, M.W., DiPietro, M., Lovett, M.C., Norman, M.K.:How Learning Works: Seven Research-Based Principles for Smart Teaching. Jossey-Bass, Hoboken (2010)
8. Anacona, J.D., Millán, E.E., Gómez, C.A.: Aplicación de los metaversos y la realidad virtual en la enseñanza. Entre Ciencia e Ingeniería **13**(25), 59–67 (2019). https://doi.org/10.31908/19098367.4015
9. Anijovich, R., González, C.: Evaluar para aprender. Conceptos e instrumentos. Aique, Buenos Aires (2021)
10. Anijovich, R., González, C.: Evaluar para aprender. Conceptos e instrumentos. Aique, Buenos Aires (2021)
11. Antonioli, M., Blake, C., Sparks, K.: Augmented reality applications in education. J. Technol. Stud. **40**(1/2), 96–107 (2014). Retrieved abril 30, 2021 from http://www.jstor.org/stable/43604312
12. Aoun, J.E.: Robot-Proof: Higher Education in the age of artificial intelligence. MIT, Massachusetts (2017)
13. Asociación Redes de Interconexión Universitaria. (n.d.).: ARIU. Retrieved enero 7, 2022 from https://riu.edu.ar/institucional/historia/
14. Asuar, B.: (2020, noviembre 11). Público. (DISPLAY CONNECTORS, SL.) Retrieved marzo 27, 2021 from https://www.publico.es/entrevistas/castells-hay-listos-establecer-ensenanza-evaluaciones-online-completo.html
15. Ausubel, D.: The use of advance organizers in the learning and retention of meaningful verbal material. J. Educ. Psychol. (51), 267–272 (1960)

U. R. Cukierman, *The Art of Teaching, the Magic of Learning: Towards Innovative Higher Education Systems*, Studies in Systems, Decision and Control 697,
https://doi.org/10.1007/978-3-032-35021-3

16. Ausubel, D.: The Psychology of Meaningful Verbal Learning. Grune & Stratton, New York (1963)
17. Ausubel, D.P., Novak, J.D., Hanesian, H.: Psicología Educativa. Un punto de vista cognoscitivo. Trillas, México (1983)
18. Autor, D.H., Salomons, A.: Is automation labor share-displacing? Productivity growth, employment, and the labor share. In: Brookings Papers on Economic Activity, pp. 1–87 (2018)
19. Babber, A.: LMS Statistics: Data, Trends & Predictions (2023). Retrieved April 1, 2023 from https://imagestation.com/lms-statistics/
20. Barton, D., Farrell, D., Mourshed, M.: Education to Employment: Designing a System that Works (2013). McKinsey & Co. Retrieved enero 14, 2022 from https://www.mckinsey.com/industries/education/our-insights/education-to-employment-designing-a-system-that-works
21. Becker, H.J.: Findings from the teaching, learning, and computing survey: Is larry Cuban right? Educ. Policy Anal. Arch. **8**(51) (2000)
22. Berners-Lee, T.: The Guardian (2020). Retrieved marzo 27, 2021 from https://www.theguardian.com/commentisfree/2020/jun/04/covid-19-internet-universal-right-lockdown-online
23. Biggs, J.: Calidad del aprendizaje universitario. Narcea, Madrid (2005)
24. Birzina, R., Pigozne, T.: Technology as a tool in STEM teaching and learning. In: Proceedings of the 13th International Scientific Conference REEP 2020. Latvia University of Life Sciences and, Jelgava (2020). https://doi.org/10.22616/REEP.2020
25. Bjork, C.: The Conversation (2023). Retrieved Noviembre 20, 2023 from https://theconversation.com/dont-fret-about-students-using-chatgpt-to-cheat-ai-is-a-bigger-threat-to-educational-equality-202842
26. Black, M.: Critical Thinking: An Introduction to Logic and Scientific Method. Prentice-Hall, New York (1952)
27. Blaxter, L., Hughes, C., Tight, M.: How to Research. Open University Press, Berkshire (2010)
28. Bloom, B.S., Engelhart, M.D., Furst, E.J., Hill, W.H., Krathwohl, D.R.: Taxonomy of Educational Objectives. The Classification of Educational Goals. Longmans, London (1956)
29. Booth, W.C., Colomb, G.G., Willi, J.M.: The Craft of Research. University of Chicago Press, Chicago (2008)
30. Bordas, M.I., Cabrera, F.M.: Estrategias de evaluación de los aprendizajes centrados en el proceso. Revista Española ed Pedagogía **LIX**(218), 25–48 (2001). Retrieved enero 25, 2024 from http://biblioteca.udgvirtual.udg.mx/jspui/bitstream/123456789/2649/1/Estrategias%20de%20evaluacion%20aprendizajes.pdf
31. Brooks, D.C., McCormack, M.: Driving Digital Transformation in Higher Education. ECAR, Louisville (2020). Retrieved enero 8, 2022 from https://library.educause.edu/-/media/files/library/2020/6/dx2020.pdf
32. Bruce, B.C., Bishop, A.P.: Using the web to support inquiry-based literacy. J. Adolescent Adult Literacy **45**(8), 706–714 (2002)
33. Bruner, J.: La educación, puerta de la cultura. Machado Grupo de Distribución, Madrid (1997)
34. Brynjolfsson, E., McAfee, A.: The Second Machine Age: Work, Progress, and Prosperity in a Time of Brilliant Technologies. W. W. Norton & Company, New York (2014)
35. Buckingham, D.: Más allá de la tecnología. Aprendizaje infantil en la era de la cultura digital. Manantial, Cambridge (2008)
36. Buckminster Fuller, R.: Ccritical Path. St. Martin's Griffin (1982)
37. Budiu, R.: Nielsen Norman Group (2017). Retrieved May 16, 2023 from https://www.nngroup.com/articles/wizards/
38. Bumstead, J.F.: The Black Board in the Primary School. A Manual for Teachers. Perkins & Marvin, Boston (1841
39. Burbules, N.C., Callister, T.A.: Educación: Riesgos y promesas de las nuevas tecnologías de la información. Granica, Buenos Aires (2001)
40. Cabero Almenara, J., Llorente Cejudo, M.: Entornos Personales de Aprendizaje (PLE): Valoración educativa a través de expertos. Areté. Revista Digital del Doctorado en Educación de la Universidad Central de Venezuela, 7–19 (2015)

41. Cabero Almenara, J., Puentes Puente, A.: La Realidad Aumentada: tecnología emergente para la sociedad del aprendizaje. Revista de Humanidades y Ciencias Sociales **66**(2), 35–51 (2020)
42. Calaprice, A.: The Ultimate Quotable Einstein. Princeton University Press, Princeton (2011)
43. Calaprice, A.: The Ultimate Quotable Einstein. Princeton University Press, Princeton (2011)
44. Calderero, J. F., Gallego, G., Meneses, G., Moreno, A.: Personalizar, un modelo para una educación de calidad en siglo XXI. Impuls Educació, Barcelona (2020)
45. Cannellotto, A.: Universidades viralizadas: la formación en y post pandemia. In: Dussel, I., Ferrante, P., Pulfer, D. (eds.), Pensar la educación en tiempos de pandemia. Entre la emergencia, el compromiso y la espera, pp. 213–228. UNIPE, Buenos Aires (2020)
46. Caratozzolo, P., Sirkis, G., Piloto, C., Correa, M.: Skills obsolescence and education global risks. In: WEEF/GEDC 2020. Ciudad del Cabo (2020). https://doi.org/10.1109/WEEF-GEDC49885.2020.9293687
47. Castañeda, L., Adell, J.: Entornos Personales de Aprendizaje: Claves para el Ecosistema Educativo en Red (2013). Marfil, Alcoy. Retrieved enero 3, 2024 from https://digitum.um.es/digitum/handle/10201/30427
48. Castells, M.: La universidad en la era de la información. Lección inaugural UOC 2017–2018. Barcelona (2017)
49. CEPAL: Tecnologías digitales para un nuevo futuro. CEPAL, Santiago (2021)
50. Chevalier Naranjo, S.: (2021, septiembre 8). Statista. From https://es.statista.com/grafico/25716/paises-y-territorios-clasificados-segun-el-indice-de-inclusion-de-internet/
51. Christina, G.: Facebook in tertiary education: the impact of social media in e-Learning. J. Univ. Teaching Learn. Pract. **17**(1), Art. 3 (2020)
52. Cobo, C.: Acepto las condiciones: Usos y abusos de las tecnologías digitales. Santillana, Madrid (2019)
53. Coicaud, S.: La democratización del conocimiento a través de la educación a distancia. Etic@net 1–13 (2002)
54. Collins, C.: Looking to the future: higher education in the metaverse. EDUCAUSE Rev. **43**(5), 50–63 (2008). Retrieved enero 26, 2022 from https://er.educause.edu/articles/2008/9/looking-to-the-future-higher-education-in-the-metaverse
55. Comenio, J.A.: Didáctica Magna. Porrúa, México (1998)
56. Comisión Europea: (2006, diciembre 18). Retrieved enero 23, 2024 from https://eur-lex.europa.eu/LexUriServ/LexUriServ.do?uri=OJ:L:2006:394:0010:0018:es:PDF
57. CONEAU: (2014). CONEAU. From https://www.coneau.gob.ar/archivos/libros_evaluacion_externa/47UTN.pdf
58. CONEAU: (2021). CONEAU. From https://www.coneau.gob.ar/coneau/nuevos-estandares-de-acreditacion-para-carreras-de-grado/
59. CONFEDI. (n.d.).: Retrieved abril 23, 2021 from https://confedi.org.ar/libro-azul-y-libro-verde/
60. CONFEDI: Marco conceptual y definición de estándares de acreditación de las carreras de ingeniería (2017). Buenos Aires, Argentina. Retrieved enero 17, 2022 from https://confedi.org.ar/wp-content/uploads/2021/07/MARCOC1.pdf
61. CONFEDI: Propuesta de Estándares de Segunda Generación para la Acreditación de Carreras de Ingeniería en la República Argentina "Libro Rojo". Universidad FASTA Ediciones (2018)
62. Congreso de la Nación Argentina: (1948, agosto 31). InfoLEG. Retrieved enero 7, 2022 from http://servicios.infoleg.gob.ar/infolegInternet/anexos/295000-299999/295701/norma.htm
63. Congreso de la Nación Argentina: (1959, octubre 14). InfoLEG. Retrieved enero 7, 2022 from http://servicios.infoleg.gob.ar/infolegInternet/anexos/195000-199999/195013/norma.htm
64. Congreso de la Nación Argentina: (1995, julio 20). InfoLEG. From http://servicios.infoleg.gob.ar/infolegInternet/anexos/25000-29999/25394/norma.htm
65. Congreso de la Nación Argentina: (2014, diciembre 16). InfoLEG. From http://servicios.infoleg.gob.ar/infolegInternet/anexos/235000-239999/239771/texact.htm
66. Congreso de la Nación Argentina: (2021, mayo 21). Diputados Argentina. Retrieved enero 8, 2022 from https://www.hcdn.gob.ar/prensa/noticias/2021/noticias_1579.html

67. Consejo Superior UTN: (2008, abril 24). Ordenanza CS N° 1182/08. Buenos Aires. Retrieved enero 27, 2022 from https://www.frba.utn.edu.ar/wp-content/uploads/2019/10/ORD-1182.pdf
68. Consejo Superior UTN: (2020, marzo 5). UTN. From https://buscadorcsu.rec.utn.edu.ar/home?documentoId=CSU**ORD**0**1753
69. Costa, A.L., Kallick, B.: Assessment Strategies for Self-Directed Learning. Corwin Press (2004)
70. Covey, S.R.: Los 7 hábitos de la gente altamente efectiva : la revolución ética en la vicia cotidiana y en la empresa. Paidos, Buenos Aires (2003)
71. Crawford, K.: Atlas de la Inteligencia Artificial. Fondo de Cultura Económica, Buenos Aires (2022)
72. Crawford, K.: Atlas de la Inteligencia Artificial. Poder, política y costos planetarios. Fondo de Cultura Económica, Buenos Aires (2022)
73. Creswell, J.W.: Educational Research. Planning, Conducting, and Evaluating Quantitative and Qualitative Research, 4th edn. Pearson Education, Boston (2012)
74. Cukierman, U.: AMERICA@UTN - learning through advanced communication and information technology resources and means @ UTN. In: 6th International Conference on Information Technology Based Higher Education and Training. ITHET, Santo Domingo (2005)
75. Cukierman, U.: Aprendizaje centrado en el estudiante: un enfoque imprescindible para la educación en ingeniería. In: Aseguramiento de la calidad y mejora de la educación en ingeniería: Experiencias en América Latina, pp. 27–39. ACOFI, Bogotá (2018)
76. Cukierman, U.R., Agüero, M., Silvestri, S., Delmonte, R., Drangosch, J., Gonzáles Corrao, L., Saclier, L., et al.: Bridging the Gap Between First-Year Students and Engineering: A Novel Application of Mobile Technologies for Improving Mathematics and Physics learning. WEEF 2017. Malaysia (2017)
77. Cukierman, U.R., Silvestri, S., Agüero, M., Drangosch, J.: A student-centered approach to learning mathematics and physics in engineering freshmen courses. In: 2018 World Engineering Education Forum. IEEExplore, Albuquerque (2018)
78. Cukierman, U., Recabarren, P.: Educación en la era de la 4ª Revolución Industrial: Competencias para un mundo donde lo único constante es el cambio. Revista Argentina de Ingeniería (RADI) **11**, 83–86 (2018)
79. Cukierman, U., Vendrell Vidal, E.: Aprendizajes reales en ambientes virtuales. El rol de la tecnología en la era de la Inteligencia Artificial y el Big Data. Cuaderno de Pedagogía Universitaria **17**(34), 59–67 (2020)
80. Cukierman, U., Vendrell Vidal, E.: Aprendizajes reales en ambientes virtuales: El rol de la tecnología en la era de la Inteligencia Artificial y el Big Data. Cuadernos de Pedagogía Universitaria **17**(34), 59–67 (2020)
81. Cukierman, U., Gonzalez, A., Iglesias, L., Ibañez, E., Rozenhauz, J., Santangelo, H., Zangara, A., et al.: Una experiencia de uso de celulares en un curso de articulaciónescuela media y universidad en modalidad a distancia. Virtual Educa Brasil (2007)
82. Cukierman, U., Perez, T., Roces, J.L.: Educación del Ingeniero del Siglo XXI. In: Saenz Quesada, M., Martí, M.L. (eds.), Problemática de la Educación en la Argentina, pp. 295–312. Academia Nacional de Ciencias Morales y Políticas, Buenos Aires (2022)
83. de Benito, B., Salinas, J.: Análisis de situaciones didácticas en los entornos virtuales de enseñanzaaprendizaje (EVEA) en la enseñanza superior. LA EDUCACIÓN EN ENTORNOS VIRTUALES: CALIDAD Y EFECTIVIDAD EN EL E-LEARNING. Edutec, Tarragona (2006)
84. Declaración de Salamanca: (2018, mayo 22). Salamanca, España. Retrieved enero 10, 2022, from https://www.crue.org/wp-content/uploads/2020/02/declaracion-de-salamanca-2018pdf.pdf
85. Dehaene, S.: ¿Cómo aprendemos? Siglo veintiuno (2019)
86. Delors, J.: Learning: The Treasure Within. UNESCO, París (1996)
87. Dennis, M.: The role of higher education in a post-COVID-19 world. Enrollment Manage. Rep. **24**(11), 3–3 (2021)

88. Design Kit. (n.d.).: Retrieved May 16, 2023 from https://www.designkit.org/methods/how-might-we.html
89. Dewey, J.: Cómo pensamos. Nueva exposición de la relación entre pensamiento reflexivo y proceso educativo. Paidos, Barcelona (1989)
90. Dirección General de Cultura y Educación de la Provincia de Buenos Aires: Sistema de Informaicón Normativa y Documental (2006). From https://normas.gba.gob.ar/documentos/x6ObL3FY.pdf
91. Donald, N.: The Psychology of Everyday Things. Basic Books, USA (1988)
92. Dondi, M., Klier, J., Panier, F., Schubert, J.: Defining the Skills Citizens Will Need in the Future World of Work. (M. &. Company, Ed.) McKinsey & Company (2021). Retrieved enero 7, 2022 from https://www.mckinsey.com/industries/public-and-social-sector/our-insights/defining-the-skills-citizens-will-need-in-the-future-world-of-work#download/%2F~%2Fmedia%2Fmckinsey%2Findustries%2Fpublic%20and%20social%20sector%2Four%20insights%2Fdefining%20the%2
93. Doran, G.T.: There's a S.M.A.R.T. way to write management's goals and objectives. Manage. Rev. **70**(11), 35–36 (1981)
94. Dougiamas, M., Taylor, P.C.: Interpretive analysis of an internet-based course constructed using a new courseware tool called Moodle. In: HERDSA 2002 Quality Conversations. Perth, Western Australia (2002)
95. Duan, H., Li, J., Fan, S., Lin, Z., Wu, X., Cai, W.: Metaverse for social good: a university campus prototype. In: Proceedings of the 29th ACM International Conference on Multimedia, pp. 153–161. ACM, New York (2021). https://doi.org/10.1145/3474085.3479238
96. Dussel, I., Ferrante, P., Pulfer, D.: Nuevas ecuaciones entre educación, sociedad, tecnología y Estado. In: Dussel, I., Ferrante, P., Pulfer, D. (eds.) Pensar la educación en tiempos de pandemia, pp. 351–364. UNIPE: Editorial Universitaria, Buenos Aires (2020)
97. EdSurge: Decoding Adaptive. Pearson, Londres (2016)
98. Educ.ar: Portal Educ.ar (2013). Retrieved Noviembre 20, 2023 from https://www.educ.ar/recursos/120640/edith-litwin-los-desafios-y-los-sinsentidos-de-las-nuevas-te
99. EDUCAUSE: 2020 EDUCAUSE Horizon Report. Teaching and Learning Edition. EDUCAUSE, Louisville (2020). From https://library.educause.edu/-/media/files/library/2020/3/2020_horizon_report_pdf.pdf
100. EDUCAUSE: 2021 EDUCAUSE Horizon Report, Information Security Edition. EDUCAUSE, Boulder (2021)
101. EDUCAUSE: 2021 EDUCAUSE Horizon Report. Teaching and Learning Edition. EDUCAUSE, Boulder (2021)
102. EDUCAUSE: 2023 EDUCAUSE Horizon Report | Teaching and Learning Edition (2023). Boulder. Retrieved Noviembre 20, 2023 from https://library.educause.edu/-/media/files/library/2023/4/2023hrteachinglearning.pdf
103. Einstein, A. (n.d.).: Quotes Cosmo. Retrieved marzo 1, 2022 from https://www.quotescosmos.com/quotes/Albert-Einstein-quote-1.html
104. Elevate Society. (n.d.).: Retrieved diciembre 29, 2023 from https://elevatesociety.com/sometimes-when-you-innovate-you/
105. Ennis, R.H.: Criticai thinking and subject specificity: clarification and needed research. Educ. Res. **18**(3), 4–10 (1989)
106. Fadel, C., Bialik, M., Trilling, B.: Educación en cuatro dimensiones. Center for Curriculum Redesign, Boston (2015)
107. Fatih, D., Charmaine, B.-K., Fahad, A.: User experience matters: does one size fit all? Evaluation of learning management systems. Technol. Knowl. Learn. **27**, 49–67 (2021)
108. Fedirko, D.: eLearning Industry (2019). From https://elearningindustry.com/digital-transformation-in-higher-education-8-top-trends
109. Felder, R.: Opinion: uses, misuses, and validity of learning styles. Adv. Eng. Educ. **8**, 1–14 (2020)
110. Felder, R.M., Brent, R.: Teaching and Learning STEM: A Practical Guide. Jossey-Bass, San Francisco (2016)

111. Felder, R.M., Brent, R.: Teaching and Learning STEM: A Practical Guide. Jossey-Bass, San Francisco (2016)
112. Felder, R.M., Silverman, L.K.: Learning and teaching styles in engineering education. Engr. Educ. **78**(7), 674–681 (1988)
113. Felder, R., Brent, R.: Active learning: an introduction. ASQ Higher Educ. Brief **2**, 4–9 (2009)
114. Feldman, P.: Digital transformation in education: from vision to practice during the pandemic. In: Plutino, A. (eds.), Languages at Work, Competent Multilinguals and the Pedagogical Challenges of COVID-19, pp. 39–45. Research-publishing.net, Voillans, Francia (2021). https://doi.org/10.14705/rpnet.2021.49.1216
115. Fernández, M.: (2020, enero 14). Infobae. From https://www.infobae.com/educacion/2020/01/14/buscaran-que-se-graduen-mas-estudiantes-en-las-areas-donde-faltan-profesionales-las-5-carreras-que-priorizarian/
116. Fleming, N.D., Mills, C.: Not another inventory, rather a catalyst for reflection. To Improve Acad. **11**, 137 (1992)
117. Fogler, H. S., LeBlanc, S.E., Rizzo, B.: Strategies for creative problem solving, 3rd edn. Pearson, Upper Saddle River, NJ (2014)
118. Forestello, R., Rivero, M., Bruni, R., Capdevila, J., Capdevila, L.: Diseño de programas de asignaturas, para el desarrollo de un aprendizaje centrado en el estudiante en las carreras de ingeniería. FCEFyN, UNC, Córdoba (2020)
119. Freire, P.: La alfabetización y el "sueño posible". Perspectivas *VI*(1), 77–81 (1976)
120. Freire, P.: Pedagogía de la autonomía. Siglo veintinuo editores, México (1997)
121. Freire, P., Faundez, A.: Por una pedagogía de la pregunta: crítica a una educación basada en respuestas a preguntas inexistentes. Siglo Veintiuno Editores Argentina, Buenos Aires (2013)
122. Frey, C.B., Osborne, M.A., Holmes, C.: TECHNOLOGY AT WORK v2.0: The Future Is Not What It Used to Be. University of Oxford, Oxford (2016)
123. Fullan, M., Scott, G.: New Pedagogies for Deep Learning Whitepaper: Education PLUS. Collaborative Impact SPC, Seattle (2014)
124. Furman, M.: Enseñar distinto: Guía para innovar sin perderse en el camino. Siglo veintiuno, Buenos Aires (2021)
125. Gamage, S., Ayres, J., Behrend, M.: A systematic review on trends in using Moodle for teaching and learning. Int. J. STEM Educ. 1–24 (2022)
126. García Aretio, L.: Los docentes: Entre tecnófilos y tecnófobos. Editorial del BENED (2007)
127. García-Peñalvo, F.J.: La percepción de la Inteligencia Artificial en contextos educativos tras el lanzamiento de ChatGPT: disrupción o pánico. Educ. Knowl. Soc. **e31279**, 24 (2023). https://doi.org/10.14201/eks.31279
128. Gardner, H.: Frames of Mind : The Theory of Multiple Intelligences. Basic Books, New York (1983)
129. Gardner, H.: The Unschooled Mind: How Children Think and How Schools Should Teach. Basic Books, New York (1991)
130. Gardner, H.: Inteligencias Multiples - La Teoria En La Practica. Paidós, Buenos Aires (2014)
131. Gardner, H., Hatch, T.: Multiple intelligences go to school: Educational implications of the theory of multiple intelligences. Educ. Res. **18**(8), 4–9 (1989)
132. Gargallo López, B., Pérez-Pérez, C., Garcia-Garcia, F., Giménez Beut, J., Portillo Poblador, N.: La competencia aprender a aprender en la universidad: propuesta de modelo teórico. Educación **XX1**, **23**(1), 19–44 (2020). https://doi.org/10.5944/educXX1.23367
133. Gartner, Inc.: Gartner (2019). From https://www.gartner.com/en/newsroom/press-releases/2019-03-26-gartner-identifies-the-top-10-strategic-technologies-
134. George Lucas Educational Foundation: Edutopia (2002). From https://www.edutopia.org/grant-wiggins-assessment
135. Gibbons, S.: Nielsen Norman Group (2016). Retrieved May 6, 2023 from https://www.nngroup.com/articles/design-thinking/
136. Glaser, B.G., Strauss, A.L.: The Discovery of Grounded Theory. Strategies for Qualitative Research. Aldine Press, Chicago (1967)
137. Google. (n.d.).: Google Académico. Retrieved enero 8, 2022 from https://scholar.google.com/

138. Google. (n.d.).: Google Scholar. Retrieved April 30, 2023 from https://scholar.google.com/
139. Graham, R.: Lograr excelencia en la formación de ingeniería: los ingredientes para un cambio exitoso. The Royal Academy of Engineering, Londres (2012)
140. Grajek, S., Reinitz, B.: EDUCAUSE (2019). Retrieved enero 8, 2022 from https://er.educause.edu/articles/2019/7/getting-ready-for-digital-transformation-change-your-culture-workforce-and-technology
141. Gray, D.: Game Storming (2009). Retrieved May 2, 2023 from https://gamestorming.com/empathy-map/
142. Greenberger, M.: Management and the Computer of the Future. John Wiley and Sons, New York (1962)
143. Grigori, D., Casati, F., Castellanos, M., Dayal, U., Sayal, M., Shan, M.: Business process intelligence. Comput. Ind. **53**, 321–343 (2004)
144. Haber, J.: Critical Thinking. The MIT Press, Cambridge (2020)
145. Hamodi, C., López Pastor, V., López Pastor, A.: Medios, técnicas e instrumentos de evaluación formativa y compartida del aprendizaje en educación superior. Perfiles educativos **37**(147), 146–161 (2015)
146. Harari: (2023a, noviembre 30). Yuval Noah Harari on GOP losses, conspiracies, AI, religion & history: Melber 'Summit Series'. (A. Melber, Interviewer) Retrieved diciembre 28, 2023 from https://youtu.be/nlni04mpDdg
147. Harari, Y.N.: 21 lecciones para el Siglo XXI. Debate, Buenos Aires (2018)
148. Harari, Y.N.: The Economist (2023). Retrieved Noviembre 19, 2023 from https://www.economist.com/by-invitation/2023/04/28/yuval-noah-harari-argues-that-ai-has-hacked-the-operating-system-of-human-civilisation
149. Harper, D. (n.d.).: Online Etymology Dictionary. Retrieved Diciembre 23, 2023 from https://www.etymonline.com/word/magic
150. Harvard Graduate School of Education. (n.d.).: Project Zero. Retrieved diciembre 25, 2023 from https://pz.harvard.edu/thinking-routines
151. Hatcher, J.: (2020, Enero 26). Inside Higher Ed. From https://www.insidehighered.com/sponsored/4-emerging-trends-will-create-opportunities-universities-2020
152. Heath, C., Heath, D.: Cambia el chip. Cómo afrontar cambios que parecen imposibles. Paidos, Mexico (2018)
153. Heaven, W.D.: The education fo ChatGPT. In: MIT Technology Review, pp. 42–47 (2023)
154. Hilliger, I., Fleet, C., Melian, C., Baier, J., Pérez-Sanagustín, M.: Offering an entrepreneurship course to all engineeringstudents: lessons learned from ING2030 in Puc-Chile. ASEE Adv. Eng. Educ. **9**(3), 1–16 (2021). Retrieved Noviembre 4, 2023 from https://www.researchgate.net/publication/354224445_Offering_an_Entrepreneurship_Course_to_All_Engineering_Students_Lessons_Learned_from_ING2030_in_Puc-Chile#fullTextFileContent
155. Hinojo Lucena, F.J., Aznar Díaz, I., Cáceres Reche, M.P., Romero Rodríguez, J.M.: Artificial intelligence in higher education: a bibliometric study on its impact in the scientific literature. Educ. Sci. **9**(1), 51 (2019)
156. Hodges, C., Moore, S., Lockee, B., Trust, T., Bond, A.: EDUCAUSE (2020). Retrieved October 20, 2020 from https://er.educause.edu/articles/2020/3/the-difference-between-emergency-remote-teaching-and-online-learning
157. Hollowell, J.: Moodle Forums (2007). Retrieved May 21, 2023 from https://moodle.org/mod/forum/discuss.php?d=80637#p357476
158. Hu, K.: Reuters (2023). Retrieved Noviembre 19, 2023 from https://www.reuters.com/technology/chatgpt-sets-record-fastest-growing-user-base-analyst-note-2023-02-01/
159. Huang, R., Spector, J.M., Yang, J.: Emerging issues in educational technology: a primer for the 21st century. In: Educational Technology, A Primer for the 21st Century, pp. 231–241. Springer, Singapur (2019). https://doi.org/10.1007/978-981-13-6643-7
160. Hughes, A.: BBC Sciende Focus (2023). Retrieved abril 26, 2023 from https://www.sciencefocus.com/future-technology/gpt-3/
161. Hui, Y.: Fragmentar el futuro. Ensayos sobre la tecnodiversidad. Caja Negra, Buenos Aires (2020)

162. IDEO. (n.d.).: Retrieved May 6, 2023 from https://designthinking.ideo.com
163. IESALC: COVID-19 and Higher Education: Impact Analysis, Policy Responses and Recommendations (2020). Retrieved October 20, 2020 from http://www.iesalc.unesco.org/en/wp-content/uploads/2020/04/COVID-19-EN-090420-2.pdf
164. Ilia, M., Shahrokh, N., Preben, H.: Exploring user experience of learning management system. Int. J. Inf. Learn. Technol. **38**(4), 344–363 (2021)
165. Inamorato dos Santos, A.: OpenEdu: las diez dimensiones del Marco europeo de educación abierta. Revista Mexicana de Bachillerato a Distancia **11**(22), 5–6 (2019)
166. Inamorato dos Santos, A., Punie, Y., Castaño-Muñoz, J.: Opening up Education: A Support Framework for Higher Education Institutions. JRC Science for Policy Report, EUR 27938 EN (2016). https://doi.org/10.2791/293408
167. INET: Demanda de capacidades 2020. INET, Buenos Aires (2016). Retrieved enero 14, 2022 from http://www.inet.edu.ar/wp-content/uploads/2016/06/2016.06.21_Informe_Demandas_Laborales_2020_vf.pdf
168. InfoLEG*:* (1995, julio 27). Retrieved diciembre 24, 2023 from https://servicios.infoleg.gob.ar/infolegInternet/anexos/25000-29999/25394/texact.htm
169. Institute of Education Sciences. (n.d.).: ERIC. Retrieved enero 8, 2022 from https://eric.ed.gov/
170. Institute of Education Sciences. (n.d.).: ERIC. Retrieved April 30, 2023 from https://eric.ed.gov/
171. Instituto Nacional de Tecnologías Educativas y de Formación del Profesorado. Ministerio de Educación y Formación Profesional. Gobierno de España: Credenciales Digitales Abiertas para el Reconocimiento de Competencias. Madrid, España (2018)
172. Interaction Design Foundation. (n.d.).: Retrieved May 16, 2023 from https://www.interaction-design.org/literature/topics/heuristic-evaluation
173. Interaction Design Foundation. (n.d.).: Retrieved May 2, 2023 from https://www.interaction-design.org/literature/topics/heuristic-evaluation
174. International Organization for Standardization: ISO Online Browsing Platform (2010). Retrieved April 8, 2023 from https://www.iso.org/obp/ui/#iso:std:iso:9241:-210:ed-1:v1:en
175. James, D.: Fichte's Social and Political Philosophy. Cambridge University Press, Cambridge (2011)
176. Jerez, O., Hasbún, B., Rittershaussen, S.: El Diseño del Syllabus en la Educación Superior. Una propuesta metodológica. Ediciones Universidad de Chile, Santiago (2015)
177. Johnson, K., Sharp, V.: Is PowerPoint crippling our students? Learn. Leading Technol. **33**(3), 6–7 (2005). Retrieved enero 13, 2022 from https://files.eric.ed.gov/fulltext/EJ728912.pdf
178. Kanwar, A., Carr, A.: The impact of COVID-19 on international higher education: new models for the new normal. J. Learn. Dev. **7**(3), 326–333 (2020)
179. Kaplan, K.: Nielsen Norman Group (2023). Retrieved May 15, 2023 from https://www.nngroup.com/articles/user-journeys-vs-user-flows/
180. Kirkwood, A., Price, L.: Technology-enhanced learning and teaching in higher education: what is 'enhanced' and how do we know? A critical literature review. Learn. Media Technol. **39**(1), 6–36 (2014). https://doi.org/10.1080/17439884.2013.770404
181. Kolb, D.A.: Learning Style Inventory Technical Manual. McBer & Co., Boston (1976)
182. Kolmos, A., Fink, F.K., Krogh, L.: The Aalborg model: problem-based and project-organized learning. In: The Aalborg Model: Progress, Diversity and Challenges, pp. 9–18. Aalborg University Press, Aalborg (2004)
183. Kotter, J.P.: Leading change: why transformation efforts fails. Harvard Bus. Rev. 59–67 (1995)
184. Kovadloff, S.: Temas de siempre. Emecé, Buenos Aires (2023)
185. Krause, S.D.: "Among the greatest benefactors of mankind": what the success of chalkboards tells us about the future of computers in the classroom. J. Midwest Modern Language Assoc. **33**(2), 6–16 (2000). Retrieved Marzo 26, 2021 from www.jstor.org/stable/1315198
186. Krause, S.D.: "Among the greatest benefactors of mankind": what the success of chalkboards tells us about the future of computers in the classroom. J. Midwest Modern Language Assoc. **33**(2), 6–16 (2000). Retrieved marzo 26, 2021 from www.jstor.org/stable/1315198

187. Layla, H.: Examining user experience of moodle e-learning system. Int. J. Adv. Comput. Sci. Appl. **12**(11), 358–366 (2021)
188. LeCun, Y., Bengio, Y., Hinton, G.: Deep learning. Nature **521**, 436–444 (2015)
189. Leigh Brown, J.: UX Booth (2018). Retrieved may 2, 2023 from https://www.uxbooth.com/articles/empathy-mapping-a-guide-to-getting-inside-a-users-head/
190. Lion, C.: Imaginar con tecnologías. Relaciones entre tecnología y conocimiento. La Crujía, Buenos Aires (2006)
191. Lion, C.: Is artificial intelligence the magic potion that will lead us to improve the quality of higher education? In: Green, E., Singh, D., Chia, R. (eds.), AI Ethics and Higher Education. Good Practice and Guidance for Educators, Learners, and Institutions, pp. 147–170. Globethics.net, Ginebra (2022)
192. Lion, C.: Is the Artificial Intelligence the magic potion that will lead us to improve the quality of Higher Education? In: Green, E., Singh, D., Chia, R. (eds.), AI Ethics and Higher Education. Good Practices and Guidance for Educators, Learners, and Institutions, p. 326. Globethics.net, Ginebra (2022)
193. Lion, C., Kap, M., Ferrarelli, M.: Universidades desafiadas: alfabetismos fluidos, hibridaciones y nuevas estrategias de enseñanza. Educación Superior y Sociedad **35**(2), 130–155 (2023)
194. Litwin, E.: ¿Cómo trabajar con tecnología en la escuela? Conferencia organizada por Fundación Telefónica y El Sabio. Buenos Aires (2002)
195. Llorens, F., Fernández, A.: Aproximación a una medida de la transformación digital de las universidades. España (2018). Retrieved abril 30, 2021 from https://www.universidadsi.es/aproximacion-a-una-medida-de-la-transformacion-digital-de-las-universidades/
196. Loh, A.P., Putra, A.S., Koh, E.C., Zuea, T.K., Tat, K.E.: ASEE—innovation, design & entrepreneurship in engineering. In: Advances in Engineering Education, pp. 1–12 (2021). Retrieved Noviembre 4, 2023 from https://advances.asee.org/wp-content/uploads/vol09/Issue3/Papers/AEE-Innovative-Loh.pdf
197. Maggio, M.: Reinventar la clase en la universidad. Paidós, Buenos Aires (2018)
198. Maggio, M.: Educación en pandemia: Guía de supervivencia para docentes y familia. Paidós, Buenos Aires (2021)
199. Maggio, M.: Híbrida: Enseñar en la universidad que no vimos venir. Tilde Editora, Buenos Aires (2023)
200. Magnani, E.: Educación y tecnologías. Adentro de la caja. In: Dusell, I., Ferrante, P., Pulfer, D. (eds.), Pensar la educación en tiempos de pandemia. Entre la emergencia, el compromiso y la espera, pp. 85–99. UNIPE, Buenos Aires (2020)
201. Malatesta, A.A.: La creación de la Universidad Obrera Nacional y la hora de la industria. La conexión universitaria entre el aula y el trabajo. Universidad de Ciencias Empresariales y Sociales, Córdoba (2010)
202. Marina, J.A.: La inteligencia fracasada: Teoría y práctica de la estupidez. Anagrama, Barcelona (2006)
203. Márquez, I.V.: Metaversos y educación: Second Life como plataforma educativa. Revisat de Comunicación y Nuevas Tecnologías **9**(2), 151–166 (2011). https://doi.org/10.7195/ri14.v9i2.30
204. Marquina, M.: Microcredenciales: el rol de las universidades en un mercado en crecimiento. Clarín (2022)
205. Martin, F., Dennen, V.P., Bonk, C.J.: A synthesis of systematic review research on emerging learning environments and technologies. Educ. Tech. Res. Dev. (68), 1613–1633 (2020). https://doi.org/10.1007/s11423-020-09812-2
206. Mason, R., Hlinka, D.: "PowerPoint" in the classroom: where is the power? Educ. Technol. **38**(5), 42–48 (1998). Retrieved enero 13, 2022 from http://www.jstor.org/stable/44428482
207. Mastache, A.: Formar personas competentes: Desarrollo de competencias tecnológicas y psicosociales. Noveduc, Buenos Aires (2007)
208. Maycock, K.W., Lambert, J., Bane, D.: Flipping learning not just content: a 4-year action research study investigating the appropriate level of flipped learning. J. Comput. Assisted Learn. (34), 661–671 (2018). https://doi.org/10.1111/jcal.12274

209. Mazur, E.: Peer instruction: getting students to think in class. In: Proceedings of ICUPE, pp. 981–988 (1997)
210. Mazur, E., Crouch, C.H.: Peer instruction: ten years of experience and results. Am. J. Phys. **69**(9), 970–977 (2001)
211. Mazza, D.: El proceso de construcción de sentido en un enfoque clínico en sentido amplio. Educación, Lenguaje y Sociedad, pp. 1–21 (2014). https://doi.org/10.19137/els-2014-111107
212. McCarthy, J., Minsky, M.L., Rochester, N., Shannon, C.: Stanford University (1955). Retrieved Noviembre 15, 2023 from http://jmc.stanford.edu/articles/dartmouth/dartmouth.pdf
213. McMillan, J.H., Hearn, J.: Student self-assessment: the key to stronger student motivation and higher achievement. Educ. Horizons **87**(1), 40–49 (2008). Retrieved enero 31, 2024 from https://files.eric.ed.gov/fulltext/EJ815370.pdf
214. McMillan, J.H., Schumacher, S.: Investigación Educativa. Pearson Educación, Madrid (2005)
215. Mei, H., Lee, C.-H., Xiang, Y.: Entrepreneurship education and students' entrepreneurial intention in higher education. Educ. Sci. **10**(9), 1–18 (2020). Retrieved Noviembre 4, 2023 from https://www.researchgate.net/publication/344513127_Entrepreneurship_Education_and_Students'_Entrepreneurial_Intention_in_Higher_Education
216. Meirieu, P.: El futuro de la Pedagogía. Teoría de la Educación, pp. 69–81 (2022). https://doi.org/10.14201/teri.27128
217. Mellul, C.: Un análisis de las tecnologías emergentes en la educación superior y en el centro de trabajo. Federación Internacional de Universidades Católicas, Paris (2018). From http://www.fiuc.org/bdf/pdf/emerging_esp.pdf
218. Microsoft: Microsoft Inclusive Design for Cognition Guidebook (2023). Retrieved May 16, 2023 from https://inclusive.microsoft.design/tools-and-activities/InclusiveDesignForCognitionGuidebook.pdf
219. Microsoft. (n.d.).: Microsoft Inclusive Design (Microsoft). Retrieved May 21, 2023 from https://inclusive.microsoft.design/
220. Milgram, P., Kishino, F.: A taxonomy of Mixed Reality visual dispays. IEICE Trans. Inf. Syst. **77**(12), 1321–1329 (1994)
221. Ministerio de Educación: Argentina.gob.ar (2020). Retrieved enero 7, 2022 from https://www.argentina.gob.ar/noticias/150-anos-de-los-primeros-ingenieros-argentinos
222. Ministerio de Educación de la Nación: (2018). Retrieved abril 9, 2021 from https://www.argentina.gob.ar/educacion/universidades/informacion/publicaciones/anuarios
223. Ministerio de Educación de la Nación: (2020, septiembre 26). From https://www.argentina.gob.ar/calidad-universitaria/plan-de-virtualizacion-de-la-educacion-superior
224. Ministerio de Educación de la Nación: (2021, septiembre 17). From https://www.argentina.gob.ar/noticias/sigamos-estudiando-una-iniciativa-que-fortalece-el-ingreso-la-universidad
225. Ministerio de Educación de la Nación: (2021, junio 7). From https://www.argentina.gob.ar/noticias/1500-millones-para-aulas-hibridas-en-universidades-de-todo-el-pais
226. Ministerio de Educación y Deportes: (2016, octubre 28). InfoLEG. From http://servicios.infoleg.gob.ar/infolegInternet/anexos/265000-269999/267158/norma.htm
227. Ministerio de Educación y Deportes de la Nación: (2017, junio 13). From https://www.coneau.gob.ar/archivos/resoluciones/RM2641_2017.pdf
228. Ministerios de Educación y de Salud de la Nación: (2020, julio 8). InfoLEG. From http://servicios.infoleg.gob.ar/infolegInternet/anexos/335000-339999/339921/norma.htm
229. Miranda, C., Goñi, J., Berhane, B., Carberry, A.: Seven challenges in conceptualizing and assessing entrepreneurial skills or mindsets in engineering entrepreneurship education. Educ. Sci. **10**(11), 1–14 (2020). https://doi.org/10.3390/educsci10110309
230. Mollick, E.: Harvard Business Review (2022). Retrieved Noviembre 19, 2023 from https://hbr.org/2022/12/chatgpt-is-a-tipping-point-for-ai
231. Mollick, E., Mollick, L.: Why All Our Classes Suddenly Became AI Classes. Must Reads. Harvard Business Publishing Education, pp. 26–32 (2023)
232. Moodle Pty Ltd. (n.d.).: Moodle. Retrieved April 8, 2023 from Moodle Pty Ltd
233. Morin, E.: Los Siete Saberes Necesarios Para la Educación del Futuro. UNESCO, París (1999)

234. Mota, J.M., Ruiz-Rube, I., Dodero, J., Figueiredo, M.: Visual environment for designing interactive learning scenarios with augmented reality. In: Proceedings of the 12th International Conference. Vilamoura (2016)
235. NCSL: School Leadership: Concepts and Evidence. University of Reading, Reading (2003)
236. Neuwirth, L., Jovic, S., Mukherji, B.R.: Reimagining higher education during and post-COVID-19: challenges and opportunities. J. Adult Continuing Educ. 1–16 (2020)
237. New Media Consortium: 2016 Horizon Report (2016). EDUCAUSE. From https://library.educause.edu/-/media/files/library/2016/2/2016hrhees.pdf
238. Nielsen, J.: Nielsen Norman Group (2020). Retrieved May 16, 2023 from https://www.nngroup.com/articles/ten-usability-heuristics/
239. Observatorio de Innovación Educativa del Tecnológico de Monterrey: Reporte Edu Trends: Aprendizaje Invertido. ITESM, México (2014). From https://observatorio.tec.mx/edutrends aprendizajeinvertido
240. OECD-Education International: Opportunities, Guidelines and Guardrails on Effective and Equitable Use of AI in Education. OECD Publishing, Paris (2023)
241. Oliver, R., Herrington, J., Reeves, T.: Authentic activities and online learning. In: Annual Conference Proceedings of Higher Education Research and Development Society of Australasia, pp. 562–567. HERDSA, Perth, Australia (2002)
242. OpenAI: ChatGPT 3.5 (2024). From https://chat.openai.com/chat
243. Oppenheimer, A.: XPosible (2020). From https://www.xposible.com/tendencias/andres-oppenheimer-es-necesario-reentrenar-a-la-gente/
244. Organización de las Naciones Unidas: Objetivos de Desarrollo Sostenible (2015). Retrieved enero 10, 2022 from https://www.un.org/sustainabledevelopment/es/education/
245. Ortega y Gasset, J.: Meditación del pueblo joven y otros ensayos sobre latinoamérica. Alianza, Madrid (1995)
246. Our World in Data: Projected World Population by Level of Education (2016). Retrieved enero 7, 2022 from https://ourworldindata.org/grapher/projection-of-world-population-ssp2-iiasa?country=~OWID_WRL
247. Oxford English Dictionary. (n.d.).: Retrieved Noviembre 15, 2023 from https://www.oed.com/
248. Oxford Martin School. University of Oxford: Our World in Data (2018). Retrieved abril 2, 2024 from https://ourworldindata.org/grapher/world-population-level-education
249. Palazzi, C.: Faro de Vigo (2015). Retrieved Noviembre 20, 2023 from https://www.farodevigo.es/salud/2015/04/02/problema-tecnologia-16932218.html
250. Páramo Morales, D.: Editorial: La teoría fundamentada (Grounded Theory), metodología cualitativa de investigación científica. Pensamiento & Gestión, vii-xiii (2015)
251. Pardo Kuklinski, H., Cobo, C.: Expandir la universidad más allá de la enseñanza remota de emergencia. Outliers School, Barcelona (2020)
252. Parser, E.: El filtro burbuja: Cómo la red decide lo que leemos y lo que pensamos. Taurus, Buenos Aires (2017)
253. Pastorino, E., Harf, R., Sarlé, P., Spinelli, A., Violante, R., Windler, R.: Programación y práctica III. Documento Curricular. PTFD. PEI. Ministerio de Educación, Buenos Aires (1995)
254. PDST. (n.d.).: Active Learning Methodologies. Retrieved diciembre 25, 2023 from https://pdst.ie/sites/default/files/teaching%20toolkit%20booklet%20without%20keyskills.pdf
255. Pedró, F.: UNESCO IESALC (2023). Retrieved enero 23, 2024 from https://www.iesalc.unesco.org/2023/12/07/la-transformacion-de-la-educacion-superior/
256. PEN: Argentina.gob.ar (2020, agosto 24). From https://www.argentina.gob.ar/noticias/el-gobierno-nacional-asegura-la-conectividad-de-toda-la-poblacion
257. Peng Chen, X.L.: A review of using Augmented Reality in Education from 2011 to 2016. In: Innovations in Smart Learning, pp. 13–18. Springer, Singapur (2016)
258. Peregrina Castaños, M.: Megapolis y ciberpolis en Snow Crash,de Neal Stephenson. Ángulo Recto. Revista de estudios sobre la ciudad como espacio plural **4**(1), 187–202 (2012). https://doi.org/10.5209/rev_ANRE.2012.v4.n1.39290

259. Pérez Serrano, G.: Investigación cualitativa. Retos, interrogantes y métodos. La Muralla, Madrid (1994)
260. Pérez, L., Miguelena, R., Diallo, A.F.: La efectividad de la formación en ambientes virtuales de aprendizaje en la educación superior. Campus Virtuales **5**(2), 10–17 (2016)
261. Perry, S.: What Is Big Data? More than Volume, Velocity and Variety… (2017). Retrieved julio 22, 2019 from https://developer.ibm.com/dwblog/2017/what-is-big-data-insight/
262. Petrocelli, H.B.: Argentina histórica (1988). Retrieved enero 7, 2022 from http://www.argentinahistorica.com.ar/intro_libros.php?tema=1&doc=57&cap=497
263. Pomerantz, J.: Learning in Three Dimensions: Report on the EDUCAUSE/HP Campus of the Future Project. EDUCAUSE, Louisville (2018)
264. Porto, S.: Banco Interamericano de Desarrollo (2018). From https://blogs.iadb.org/conocimiento-abierto/es/como-las-insignias-abiertas-digitales-pueden-ayudar-destacar-competencias-profesionales-2/
265. Prabook. (n.d.).: Prabook. Retrieved May 21, 2023 from https://prabook.com/web/martin.dougiamas/2274527
266. PUC de Chile. (n.d.).: Actividades de Aprendizaje Activo. Retrieved diciembre 25, 2023 from https://desarrollodocente.uc.cl/recursos/tematicas-docentes/aprendizaje-activo/estrategias-de-aprendizaje-activo/
267. Querci della Rovere, E.: LinkedIn (2023). Retrieved Noviembre 19, 2023 from https://www.linkedin.com/posts/edoardoquercidellarovere_spesso-mi-capita-di-leggere-e-sentire-conversazioni-activity-7126095949738721280-amQI?utm_source=share&utm_medium=member_desktop
268. Rahman, S.F., Yunus, M.M., Hashim, H.: The uniqueness of flipped learning approach. Int. J. Educ. Pract. **8**(3), 394–404 (2020). https://doi.org/10.18488/journal.61.2020.83.394.404
269. Ramaswamy, S.: Nielsen Norman Group (2023). Retrieved May 2, 2023 from https://www.nngroup.com/articles/data-findings-insights-differences/
270. RANDSTAD ARGENTINA SA: (2020, enero 8). Retrieved abril 9, 2021 from https://www.randstad.com.ar/quienes-somos/press-releases/randstad-identifica-5-sectores-con-demanda-laboral-activa-2020/
271. Real Academia Española. (n.d.).: Diccionario de la lengua española. Retrieved Diciembre 19, 2023 from https://dle.rae.es/arte
272. Resolviendo la Escasez de Talento: (2018). Manpower Group. Retrieved enero 14, 2022 from https://www.manpowergroup.com.ar/wps/wcm/connect/manpowergroup/ced492e5-ffa1-4538-9192-613ceeda22f4/Encuesta+de+Escasez+de+Talento+2018.pdf
273. Reynders, G., Lantz, J., Ruder, S.M., Stanford, C.L., Cole, R.S.: Rubrics to assess critical thinking and information processing in undergraduate STEM courses. Int. J. STEM Educ. **7**(9), 1–15 (2020)
274. Rigal, L., Sirvent, M.: La naturaleza de la investigación científica de lo social. In: Rigal, L., Sirvent, M. (eds.), Metodología de la Investigación social y educativa: diferentes caminos de producción de conocimiento (en elaboración), pp. 1–20. Buenos Aires (2007)
275. Rivas, A.: Cambio e innovación educativa : las cuestiones cruciales : documento básico, XII Foro Latinoamericano de Educación. Fundación Santillana, Buenos Aires (2017)
276. Rivas, A.: The Platformization of Education: A framework to Map the New Directions of Hybrid Education Systems. UNESCO, Ginebra (2021). International Bureau of Education. Retrieved enero 8, 2022 from https://unesdoc.unesco.org/ark:/48223/pf0000377733
277. Robinson, K.: The Element. How Finding Your Passion Changes Everything. Penguin, Londres (2009)
278. Robinson, K.: TED Talks (2013). Retrieved enero 24, 2024 from https://www.ted.com/talks/sir_ken_robinson_how_to_escape_education_s_death_valley
279. Robinson, K., Aronica, L.: Escuelas creativas. La revolución que está transformando la educación. Grijalbo, Barcelona (2015)
280. Rodriguez-Gomez, D., Castro, D., Meneses, J.: Problematic uses of ICT among young people in their personal and school life. Comunicar: Media Educ. Res. J. **XXVI**(56), 91–100 (2018)

281. Rolheiser, C., Ross, J.A.: Student self-evaluation: what research says and what practice shows. In: Small, R.D., Thomas, A. (eds.), Plain Talk about Kids, pp. 43–57. Center for Development and Learning, Covington, LA (2001)
282. Rolón, G.: La felicidad. Más allá de la ilusión. Planeta, Buenos Aires (2023)
283. Royal Academy of Engineering: Global Engineering Capability Review. Royal Academy of Engineering, Londres (2020). Retrieved enero 7, 2022 from https://www.raeng.org.uk/publications/reports/global-engineering-capability-review
284. Sábato, J.A.: Ensayos en campera. Juarez Editor, Buenos Aires (1979)
285. Sabino, C.: El proceso de investigación. Panapo, Caracas (1992)
286. Sala, A., Punie, Y., Garkov, V., Cabrera, M.: LifeComp The European Framework for Personal, Social and Learning to Learn Key Competence. Office of the European Union, Luxemburgo (2020)
287. Samuel, A.L.: Some studies in machine learning using the game of checkers. IBM J. Res. Dev. **3**(3), 210–229 (1959)
288. Sandle, T.: Digital Journal (2018). Retrieved febrero 9, 2024 from https://www.digitaljournal.com/tech-science/op-ed-knowledge-doubles-almost-every-day-and-it-s-set-to-increase/article/537543
289. Schein, E.H.: The Corporate Culture Survival Guide. Jossey Bass (2009)
290. Schor, D., Lim, T.J., Kinsner, W.: The future of engineering education. In: IEEE Potentials, pp. 5–6 (2021)
291. Secretaría de Políticas Universitarias: Impacto de la pandemia COVID-19 en las rutinas educativas de las universidades. Buenos Aires (2020). Retrieved agosto 6, 2021 from http://www.unq.edu.ar/noticias/5066-impacto-de-la-pandemia-covid-19-en-las-rutinas-educativas-de-las-universidades.php
292. Secretaría de Políticas Universitarias. (n.d.).: Sistema de consulta de estadísticas universitarias. Retrieved enero 7, 2022 from https://estadisticasuniversitarias.me.gov.ar/#/home/2
293. Sigman, M., Bilinkis, S.: Artificial. La nueva inteligencia y el contorno de lo humano. DEBATE, Buenos Aires (2023)
294. Sigman, M., Bilinkis, S.: Artificial. La nueva inteligencia y el contorno de lo humano. DEBATE, Buenos Aires (2023)
295. Sosa Neira, E.A., Salinas, J., de Benito Crosetti, B.: Emerging technologies (ETs) in education: a systematic review of the literature published between 2006 and 2016. Int. J. Emerging Technol. Learn. **12**(5), 128–149 (2017). https://doi.org/10.3991/ijet.v12i05.6939
296. Stake, R.: Investigación con estudio de casos (2ds ed.). Morata, Madrid (1999)
297. Stallings, W.: Data and Computer Communications. Pearson, New Jersey (2014)
298. Stephenson, N.: Snow Crash. Bantam Books, New York (1992)
299. Tait, A.: Reflections on student support in open and distance learning. Int. Rev. Res. Open Distance Learn. **4**(1), 1–9 (2003)
300. Talanquer, V., Bucat, R., Tasker, R., Mahaffy, P.: Lessons from a pandemic: educating for complexity, change, uncertainty, vulnerability, and resilience. J. Chem. Educ. **97**(9), 2696–2700 (2020)
301. Talbert, R.: Flipped Learning. A Guide for Higher Education Faculty. Stylus, Sterling (2017)
302. Tania, V.: TNW (2020). Retrieved April 30, 2023 from https://thenextweb.com/news/a-brief-history-of-ux-design-and-its-evolution
303. TED: TED Blog (2009). Retrieved enero 24, 2024 from https://blog.ted.com/ted_and_reddit_1/
304. Terragno, R.: La Argentina del Siglo 21. Sudamericana Planeta, Buenos Aires (1985)
305. The Foundation for Critical Thinking. (n.d.).: Retrieved febrero 12, 2024 from https://www.criticalthinking.org/pages/our-conception-of-critical-thinking/411
306. Tiulkanov, A.: LinkedIn (2023). From https://www.linkedin.com/posts/tyulkanov_a-simple-algorithm-to-decide-whether-to-use-activity-7021766139605078016-x8Q9?utm_source=share&utm_medium=member_desktop
307. Tobón, S.: Ejes esenciales de la sociedad del conocimiento y la socioformación. Kresearch, Mount Dora (2017)

308. Torres Kompen, R., Edirisingha, P., Canaleta, X., Alsina, M., Monguet, J.: Personal learning environments based on web 2.0 services in higher education. Telematics Informatics **38**, 194–206 (2019). https://doi.org/10.1016/j.tele.2018.10.003
309. Torrico, B.: Puntos sobre la i (Banco Interamericano de Desarrollo) (2021). From https://blogs.iadb.org/innovacion/es/digitalizacion-acelerada-lo-que-la-pandemia-le-enseno-a-las-universidades/
310. Turing, A.M.: Computinh machinery and intelligence. Mind **49**, 433–460 (1950)
311. UNESCO: UNESDOC Digital Library (2016). Retrieved enero 10, 2022, from https://unesdoc.unesco.org/ark:/48223/pf0000245656_spa
312. UNESCO: ¿Cerrar ahora para reabrir mejor mañana? La continuidad pedagógica en las universidades de América Latina durante la pandemia. UNESCO, París (2021a). Retrieved agosto 13, 2021 from https://www.iesalc.unesco.org/wp-content/uploads/2021/07/Cerrar-ahora-para-reabrir-mejor-manana-FINAL-1.pdf
313. UNESCO: COVID-19: Reopening and Reimagining Universities, Survey on Higher Education Through the UNESCO National Commissions. UNESCO, París (2021b). Retrieved agosto 13, 2021 from https://unesdoc.unesco.org/ark:/48223/pf0000378174
314. UNESCO: Engineering for Sustainable Development. UNESCO, Beijing (2021c). Retrieved enero 7, 2022 from https://unesdoc.unesco.org/ark:/48223/pf0000375644.locale=en
315. UNESCO: ChatGPT and Artificial Inteligence in Higher Education. Quick Start Guide. UNESCO, París (2023)
316. UNESCO: ChatGPT and Artificial Intelligence in Higher Education. Caracas (2023)
317. UNESCO IESALC: Pensar más allá de los límites. Perspectivas sobre los futuros de la educación superior hasta 2050. UNESCO, París (2021). Retrieved enero 8, 2022 from https://www.iesalc.unesco.org/los-futuros-de-la-educacion-superior/pensando-mas-alla-de-los-limites-perspectivas-sobre-los-futuros-de-la-educacion-superior-hasta-2050/
318. UNESCO Institute for Lifelong Learning: Harnessing the Potential of ICTs. UNESCO, Hamburg (2016)
319. Universidad Tecnológica Nacional: Repensando la UTN 2030 (2020). Retrieved agosto 6, 2021 from https://www.utn.edu.ar/es/articulos-slider-principal/la-utn-publica-el-informe-repensando-utn-2030
320. UNRC: UNRC (2017, julio 31). Retrieved from https://www.unrc.edu.ar/unrc/academica/docs/piimei/lin-orientar-innovacion-curricular.pdf
321. UNS: Noticias Universitarias (2021, diciembre 14). From https://st02.uns.edu.ar/noticias/documentos/220.docx
322. UTN–FRBA: (2021, agosto 27). From https://www.frba.utn.edu.ar/la-facultad-aprobo-el-proyecto-educativo-mixto-en-la-formacion-en-ingenieria/
323. UTN: (2008, marzo 13). From http://csu.rec.utn.edu.ar/docs/php/salida_nuevo_sitio_rectorado.php3?tipo=RES&numero=133&anio=2008&facultad=CSU
324. UTN: (2011, septiembre 14). Estatuto. Retrieved enero 18, 2022 from http://csu.rec.utn.edu.ar/AU/RES/2011/1.pdf
325. UTN: (2020, marzo 5). Ordenanza CS N° 1753/20. Buenos Aires, Argentina. Retrieved enero 20, 2022 from http://csu.rec.utn.edu.ar/CSU/ORD/1753.pdf
326. Vial, G.: Understanding digital transformation: a review and a research agenda. J. Strategic Inf. Syst. **28**(2), 118–144 (2019). https://doi.org/10.1016/j.jsis.2019.01.003
327. Vilchis, N.: Instituto para el Futuro de la Educación. Tecnológico de Monterrey (2024). From https://observatorio.tec.mx/edu-news/habilidades-digitales-futuro/
328. Walker, R.: La realización de estudios de casos en educación. Ética, teoría y procedimientos. In: Dockrell, W.B., Hamilton, D., (comps.), Nuevas reflexiones sobre la investigación educativa, pp. 42–82. Narcea, Madrid (1983)
329. Walter, Y.: Embracing the future of Artificial Intelligence in the classroom: the relevance of AI literacy, prompt engineering, and critical thinking in modern education. Int. J. Educ. Technol. Higher Educ. **21**(15), 1–29 (2024)
330. Web Analytics Association: Web Analytics Definitions (2007). Retrieved julio 23, 2019 from http://www.inesting.org/ad2006/adminsc1/app/marketingtecnologico/uploads/Manuais/waa-standards-analytics-definitions-volume-i-20070816.pdf

331. Wiggins, G.: A true test: toward more authentic and equitable assessment. Phi Delta Kappan **7**(9), 703–713 (1989)
332. Wiggins, G., McTighe, J.: Understanding by Design. ASCD, Virginia, USA (2005)
333. Wong, W.: EdTech (2020, Febrero 13). From https://edtechmagazine.com/higher/article/2020/02/it-trends-watch-higher-education-moves-new-decade
334. World Economic Forum: The Future of Jobs Report. WEF (2020). Retrieved enero 7, 2022 from https://www3.weforum.org/docs/WEF_Future_of_Jobs_2020.pdf
335. World Economic Forum: Future of Jobs Report 2023. WEF, Geneva, Switzerland (2023)
336. Yale university. (n.d.).: Usability & Web Accessibility. Retrieved May 2, 2023 from https://usability.yale.edu/understanding-your-user/user-journey-maps
337. Yániz, C.: Las competencias en el currículo universitario: implicaciones para diseñar el aprendizaje y para la formación del profesorado. Red U. Revista de Docencia Universitaria, pp. 1–13 (2008)
338. Yin, R.: Investigación sobre estudio de casos: Diseños y métodos. Sage Publications,Thousand Oaks (1994)

GPSR Compliance

The European Union's (EU) General Product Safety Regulation (GPSR) is a set of rules that requires consumer products to be safe and our obligations to ensure this.

If you have any concerns about our products, you can contact us on ProductSafety@springernature.com

In case Publisher is established outside the EU, the EU authorized representative is:

Springer Nature Customer Service Center GmbH
Europaplatz 3
69115 Heidelberg, Germany

Batch number: 10809552

Printed by Printforce, the Netherlands